Office Technology

INFORMATION PROCESSING SKILLS:

MATHEMATICS

Thomas G. Sticht
Barbara A. McDonald

GOALS
Glencoe Occupational Adult Learning Series

GLENCOE
Macmillan/McGraw-Hill

New York, New York   Columbus, Ohio   Mission Hills, California   Peoria, Illinois

This program was prepared with the assistance of
Chestnut Hill Enterprises, Inc.

Send all inquiries to:

GLENCOE DIVISION
Macmillan/McGraw-Hill
936 Eastwind Drive
Westerville, OH 43081

ISBN 0-07-061515-2

2 3 4 5 6 7 8 9 0   POH   99 98 97 96 95 94 93 92

Table of Contents

Introduction

As an employee in an office, you will need to do different kinds of math. *Information Processing Skills: Mathematics* will teach you how to do these tasks. Use this book with the *Office Technology Knowledge Base*.

You may work in an office as an entry-level worker, in an advanced position, or as a manager. In all of these jobs, you need to use math skills in three areas.

- ❏ Math helps you to handle **financial resources** on the job. One task in this area is keeping correct checkbook records. This includes recording checks and deposits and checking account balances. Another task is figuring the company payroll and keeping records of what people have been paid.

- ❏ Math is also needed in working with **material resources** on the job. You use math in handling shipping for the company. Math is also needed for making copies of documents and for travel arrangements.

- ❏ The third area where math is used is in handling **human resources.** Human resources are the people in a company. You use math in developing statistics and in producing graphs about the company's employees.

In Part 1 of this book, you will try some math calculations. You will find out how much you recall about basic math skills, such as adding, subtracting, multiplying, dividing, and working with fractions and decimals. You will see which skills you need to work on. Part 7 of this book has a review of all the important math skills. You can use Part 7 to brush up on your math.

When you check your answers for Part 1, write the results in the table on page 183. This table will show you which skills you should review. It will tell you which pages in Part 7 you should study to improve these skills.

After you have refreshed your math skills in Part 7, you will be ready to proceed through the rest of this book.

In Part 2, you will learn information processing skills that will make math tasks easier for you. Part 3 shows you how to use these skills as you do the math tasks in this book. Parts 4, 5, and 6 show you how to do math tasks in the areas of financial resources, material resources, and human resources. You will use information from the *Office Technology Knowledge Base* as you work on these tasks.

PART 1

PERFORMING MATHEMATICS CALCULATIONS

In this part of the book, you will do mathematics calculations. This is a chance for you to try some basic calculations to see how much you remember. If you have trouble with the calculations, you can refresh your skills by using Part 7 of this book. But first, try to do the following problems:

1. At the Gibson Paper Company, you have records of the number of photocopies the Copy Center made for one of your customers. The records show these numbers of copies made on different days: 45, 150, 226, 1,010, 31. How many copies were made in all? (Add the numbers.)

2. You work in the Copy Center at Gibson. You know that there were 5,500 pieces of photocopy paper in stock at the beginning of the month. There are 825 pieces left. How many pieces were used? (Subtract the numbers.)

3. The time sheets for the office show that 16 people each worked about 45 hours on a project. Your supervisor wants to know about how many hours were spent on the project. (Multiply to get an estimate.)

4. You want to see how many pages the Desktop Publishing Department at Gibson can produce in an hour. Over the last week the department worked 35 hours and produced 525 pages. (Divide to find out how many pages the department produced per hour.)

5. You are working on word processing a document. You need to see how much space to allow in a document for two diagrams. The lengths of the diagrams are $1\frac{3}{4}$ inches and $3\frac{3}{8}$ inches. (Add the lengths to find the total length.)

6. You are producing a report for your supervisor. A piece of paper needs to fit into a folder that is $6\frac{1}{2}$ inches wide. The paper is $9\frac{1}{5}$ inches wide. How much do you have to cut from the paper to make it fit? (Subtract the numbers.)

7. As an office assistant, you are selecting new furniture for your work space. You have a space $7\frac{1}{2}$ feet wide, and you can use $\frac{1}{3}$ of it for a computer table. How wide can the table be? (Multiply. Write your answer in lowest terms.)

8. You are preparing a report on the desktop publishing system. There are $6\frac{1}{2}$ inches of space on a piece of paper for columns of

figures. Each column should be ½ an inch wide. How many columns will fit in the space? (Divide 6½ by ½.)

9. You are working in Gibson's Mailroom and have a group of letters needing first-class postage. The letters weigh 1 ounce, 3 ounces, 7 ounces, and 11 ounces. Look at the table in Figure 3-3 in Chapter 3 of the *Office Technology Knowledge Base*. How much postage is needed for each letter? How much will the postage cost in all? (Add.)

10. You are buying office furniture. You see that a desk lamp costs $59.25. To figure the sales tax on the lamp, multiply by 0.08. What is the sales tax? (Multiply.)

11. You are a clerk in the Accounting Department at Gibson. You have the check register shown in Figure 9-3 of Chapter 9 of the *Office Technology Knowledge Base*. The next check you write is to the Water Department for $176.45. (Subtract that amount from the balance to find the new balance in the checking account.)

12. The office time cards show that a word processing temporary employee puts in 7.25 hours per day. Of that time, the employee spends 0.75 hours learning the office methods. What portion of time is that? (Divide 0.75 by 7.25. Round your answer to hundredths.)

13. Two printer ribbons were used to print 1,600 pages. How many pages can you print with 3 ribbons? (Write and solve a proportion.)

14. The Copy Center started with a page that was 16 inches wide. It reduced the photo by 4 inches. What percentage of 16 is 4? (Write and solve a problem using percents.)

15. Every year Gibson pays for service on its photocopiers. Last year the cost for the service was $3,000. This year the company tells you the cost will be $3,450. What percent increase is this?

16. Legal pads usually cost $18.50 a dozen. The office supply store has them on sale for a 20 percent discount. What is the new price?

17. Figure 6-11 in Chapter 6 of the *Office Technology Knowledge Base* shows a portion of a database for the company payroll. Look in the column labeled "Wage" to answer these questions.
    - ❏ What is the mean of the wages paid to these employees?
    - ❏ What is the median wage paid?
    - ❏ What is the mode for these wages?

18. You plan to put shelves around the perimeter of a rectangular storage room. If the sides of the rectangle are 12 feet and 6 feet, how long is the perimeter of the room?

19. Joseph Netti's office in the Accounting Department needs new carpeting. The room is a square with 18 feet on a side. How many square feet of carpeting will be needed to cover the floor?

20. You are asked to figure how large an air conditioner to buy for the Reception Area. You need to know the volume of the room. Think of the room as a rectangular prism. It is 9 feet high, 25 feet long, and 19 feet wide. What is the volume of the room in cubic feet?

PART 2

OVERVIEW OF THE THREE Cs

❏ THE THREE Cs: COMPREHENSION, COMPUTATION, COMMUNICATION

Probably when you learned mathematics in school, you spent most of your time learning and practicing basic mathematics skills. You added, subtracted, multiplied, and divided with different kinds of numbers. You solved a lot of math problems where someone else gave you the numbers to work with. When you do mathematics in a work setting, however, no one gives you math problems to solve. Instead, you have to:

❏ Decide what task you need to do and how mathematics can help you do it. You need to put together the numbers to work with.

❏ Decide what computation you need to do and then do it. This is where you use your basic math skills.

❏ Communicate to someone else about the results of your work.

Three kinds of mathematics information processing skills are needed to achieve these tasks. These are the "3 Cs":

❏ **Comprehend** what the task is.

❏ **Compute**.

❏ **Communicate** to yourself and to others about the work and the answers.

COMPREHEND

Comprehend is the first of the 3 Cs. In order to do mathematics, you must first **comprehend** the problem. Comprehend means understand. When you understand a problem, you can solve it. You can compare solving a mathematics problem to driving a car. Before you drive your car, you must know where you need to go. Then you can decide on the route to take and drive there. When you do a math problem, you must know first what you want to accomplish. Then you can decide how to do it. Often people see the numbers in a math problem and immediately add, subtract, multiply, or divide them. This is like jumping in the car and starting to drive without thinking about where you want to end up.

To comprehend a math problem, you must read the problem and understand what it says. You can improve your understanding by using a few reading strategies. Read the problem over two or more times. Try saying the problem in your own words. Then:

❏ State what you are to do. To help yourself, think about what question is being asked. What information is missing? What do you need to find?

❏ Decide what steps to take to solve the problem. Think about how you will gather the necessary information. Remember formulas and rules that apply to this kind of situation. It may help you to think about similar problems that you have solved in the past. How did you solve them?

❏ Find the information you need to solve the problem. Look in the problem itself to see what facts are given. Ignore information that you don't need.

❏ Decide what system of measurement you are using. In your math textbooks, you worked mainly on numbers. In the real world, you always work in some system of measurement. You work with numbers that stand for *time, money, length, weight, temperature, area, volume,* or even *people.* These are the things that are meaningful on the job.

Taking these steps will help you comprehend your task. Once you comprehend, it will be easier to solve the problem.

COMPUTE

The second *C* in the 3 Cs is **compute**. After you comprehend a problem, you'll need to decide what kind of computation to do. Then you can perform the necessary computation. Figure 2-1 below shows the kinds of choices you make when you decide how to compute.

The Mathematics Knowledge Base

Basic Operations	**Types of Numbers**	**Systems of Measure**
Addition	Whole Numbers	U.S. Standard Measures
Subtraction	Fractions	Metric
Multiplication	Decimals	Length
Division	"Mixed"	Area
	Signed (+;-)	Weight
		Volume
		Time
		Temperature
		Money

Relationships Between Numbers	**Statistics and Graphs**
Percentages	Count Distributions
Ratios	Central Tendency
Proportions	Variability (Range)
	Bar Graph
	Line Graph
	Circle Graph

Figure 2-1

There are only four types of basic operations: addition, subtraction, multiplication, and division. These are shown in the column at the left.

You choose from among these operations. Sometimes, you'll need to do more than one operation. You may need to add the cost of several checks and then subtract the total from the balance in the checking account, for example.

Sometimes instead of performing operations with numbers, you use numbers to make comparisons. Ratios, percentages, and proportions all describe relationships between quantities. They are shown below the basic operations in Figure 2-1. When you want to see what part of an employee's salary goes for taxes and what part is take-home pay, for example, you show the two parts as percentages.

You perform the basic operations or comparisons on one of several types of numbers. These are shown in the center column at the top. Types of numbers include whole numbers, fractions, decimals, mixed numbers (whole numbers and fractions such as $1\frac{1}{2}$ or $35\frac{5}{8}$), and signed numbers (positive and negative numbers shown with + or - signs). You can add, subtract, multiply, or divide with:

whole numbers: $24 \div 6$ $45{,}768 \times 82$

fractions: $\frac{1}{4} \times \frac{3}{7}$ $\frac{1}{2} + \frac{5}{6}$

decimals: $0.15 \div 0.8$ $\frac{12}{567} \times 0.01$

mixed numbers: $1\frac{4}{5} + 6\frac{7}{8}$ $66\frac{1}{2} \times 8$

signed numbers: $^-8 + {}^+12$ $^+45 \times {}^-19$

The numbers you work with stand for units in a system of measurement. You must always decide what system of measurement to use. Some systems of measurement are shown in the column at the right in Figure 2-1. You might, for example, do the basic operation of addition on fractions that stand for time. Your answer will be a length of time.

Hours worked: $1\frac{1}{4}$ $6\frac{1}{2}$ $5\frac{3}{4}$

System of
Measurement
↓

Total hours $= 1\frac{1}{4} + 6\frac{1}{2} + 5\frac{3}{4} = 13\frac{1}{2}$ **hours**

Another part of the mathematics knowledge base is statistics. Statistics are facts in number form. You are probably familiar with the common statistic called the "mean" or "average." There are other statistics as well. Some of them are shown in the column at the bottom center in Figure 2-1. Number facts can also be presented in graphs. Some common graphs are named in the figure. You may choose to use statistics or graphs to solve a problem.

Once you have decided what computation to do, you do the work. If the computation involves several steps and several operations, you need to be sure to do them all.

COMMUNICATE

The third *C* in the 3 Cs is **communicate**. Communication is very important in any job situation. It's just as important to communicate about mathematics tasks on the job as it is to communicate about things you have read. When you solve a math problem on the job, you do it because someone needs the information. You solve a problem about a payroll so that someone knows what numbers to put on a paycheck. You solve a problem about the cost of a business trip so that your supervisor can decide whether or not to spend the money on the trip. You find the checking account balance so that your supervisor knows how much money the company has in the bank.

Sometimes you are the person who needs the information. You find the cost of mailing a package so that you can put the correct postage on the package.

Usually you communicate the results of your math work in writing. It's very easy to forget numbers unless they are written down. So you write a note to yourself or your supervisor about the results of your work. Or you write the results in a log, a register, or a chart. You might prepare a graph to show your results.

❏ HOW THE 3 CS WILL HELP YOU

Using the 3 Cs to solve math problems will help you to get the correct solutions to the problems. It will help you to apply the math skills you have to real-world math problems on the job. You'll have an orderly approach to use each time you encounter a math problem. You may find that you understand what you are doing in mathematics for the first time.

Figure 2-2 summarizes the 3 Cs.

---

1 — Comprehend
- ❏ State what you are to do.
- ❏ Decide on the steps you should follow.
- ❏ Collect the necessary information.
- ❏ Decide what system of measurement to use.

2 — Compute
- ❏ Decide what computation to do. This involves deciding on the operation to do, the types of numbers you need to use, the system of measurement, and whether to use statistics or graphs as part of the process.
- ❏ Do the computation.

3 — Communicate
- ❏ Communicate the results to yourself and others in writing.

Figure 2-2

3

HOW TO USE THIS BOOK

This book is designed to help you to do mathematics on the job. It is divided into three kinds of mathematics—mathematics for financial resources, mathematics for material resources, and mathematics for human resources. You will learn about each of these three types of job mathematics using the 3 Cs.

All the mathematics in this book is taught in a way that will be useful for your career in an office. Look ahead now at Parts 4, 5, and 6. You'll see that each contains several job situations. You'll learn about the kind of mathematics that you'll need in these job situations. For example, you'll learn about the mathematics used in computing postage and shipping costs. This will be useful to you if you have a job in a mailroom. You'll learn about figuring payroll deductions and take-home pay. This will be useful to you if you have a job as a bookkeeping clerk. You'll learn about figuring car rental costs. This will be useful to you if you have a job as an office assistant.

Here's how to use Parts 4, 5, and 6.

❏ First, read about the job situation. This section will tell you something about one of the jobs in an office and about some of the work you would do in that job. This section may tell you that some information is found in the *Office Technology Knowledge Base*. For example, the section may tell you to look at a figure in the *Knowledge Base*. Look at the *Knowledge Base* when you are instructed to do so.

The job situation will not ask you to do any mathematics. It will give you an idea of how mathematics is used in this job situation. For example, look ahead to the start of Part 4. The first situation is "Checking Accounts." The text there tells you about how a book-keeper works with checking accounts.

❏ Next you will see the heading "Task 1." This section will describe one kind of mathematics work in the job situation. Read this section carefully. Pay special attention to any formulas or rules.

Look ahead to the start of Part 4. On page 10, you will see that the first task in this job situation is "Task 1 Checkbook Register—Recording Payments." This section tells you how to record a check and gives you the formula for finding the new balance.

❏ Under each task, you will find a section called "Using the 3 Cs." This section shows you how to solve a math problem called for by the job task. The section gives you a sample problem.

On page 10, you'll see this math problem.

You write a check to Northern Electric for the monthly electric bill. The date is January 10. The check number is 7800, and the amount of the check is $347.50. Record the information in the register. Calculate the new balance.

Below the math problem, you will see the headings "Comprehend," "Compute," and "Communicate." These are the 3 Cs. Under each heading is a set of steps and the answer you would write if you were solving the problem.

Here is what you'd see under "Comprehend":

COMPREHEND
In the space below, write what you are to do.

Record the check in the register and find the new balance.

Write the steps you should follow to solve the problem.

1. Record all the information about the current check.
2. Subtract the new check amount from the previous balance.

Write the information you need to solve the problem.

Check information: Northern Electric, check number 7800, date January 10, amount of the check $347.50, previous balance $5,500.00.

Write the system of measurement.

Money

Read the problem and read each step. Try to think of the answer you would give before you read the answer in the book.

❑ After the example, you'll see the heading "Practice." Following this heading are a number of math problems for you to do. They will be like the one worked out in "Using the 3 Cs." Read each problem, and write your answer for each step of the 3 Cs. Look back at the example in "Using the 3 Cs" for help on completing each step. You will need to refer to information in the *Office Technology Knowledge Base* to complete some tasks.

Answers to each problem are found in the Answer Key at the back of the book. Check your progress by comparing your answers to the answers there. You may not have used the same words shown in the answers. This may be all right, if your answer has the same meaning. Check with your instructor.

❑ After you complete all the math problems for the first task, you will go on to do other tasks for the same job situation. Each one will have information on the task, an example under "Using the 3 Cs," and a set of "Practice" exercises.

Check with your instructor to see when you should read the *Knowledge Base* and when you should begin work on Parts 4, 5, and 6.

PART 4
MATHEMATICS AND FINANCIAL RESOURCES

Part 4 covers some of the financial resources you may deal with on the job. There are two job situations in this part of the book: checking accounts and payroll. Each job situation requires you to do several math tasks. Use the 3 Cs to complete each math task.

❏ JOB SITUATION 1 CHECKING ACCOUNTS

Chapter 9 of the *Office Technology Knowledge Base* discusses checking accounts. Businesses use checking accounts to pay bills. As a clerk in the Gibson Paper Accounting Department, you often write checks to pay bills. You need to keep track of checks written.

When you write a check, you write the name of the payee (the person or company that you are paying). You write the amount in numbers and words. The check will be signed by an officer of the company. You also record the payment, on either a check stub or in a check register. Both the check stub and the register include the same information—the date, check number, amount of check, payee's name, and new balance in the account.

As a bookkeeper or accounting clerk, you also take checks that other companies have sent to Gibson and deposit them in Gibson's bank account. You need to record these deposits as well. You write the date, amount of deposit, and the source of the money (whom or what company it is from). You figure the new balance in the account.

Finally, you reconcile the checking account once a month against the bank's own records. The bank sends the company a statement showing every check cashed and every deposit made during the month. The bank's records and the company's check register must agree. Figure 9-3 in your *Knowledge Base* shows a check register.

TASK 1: CHECKBOOK REGISTER—RECORDING PAYMENTS

You are a bookkeeper at Gibson Paper Company. Like Joseph Nitti, one of your jobs is writing checks to pay bills. You record each payment in the checking account register. Look at the example in Figure 9-3 of the *Knowledge Base*. For each check, you must:

1. Record all the information about the current check—the check number, the date, the payee's name, and the amount of the check.

2. Find the new checking account balance—the amount of money now in the account. Use this formula:

    previous balance – amount of check = new balance

    The previous balance is always the balance shown in the column at the far right just above your check. In the sample checkbook register shown in the *Knowledge Base*, the balance on January 2 was $4,250; on January 5, it was $5,500.

USING THE 3 CS

You write a check to Northern Electric for the monthly electric bill. The date is January 10. The check number is 7800, and the amount of the check is $347.50. Record the information in the register. Calculate the new balance.

COMPREHEND

In the space below, write what you are to do.

Record the check in the register and find the new balance.

Write the steps you should follow to solve the problem.

1. Record all the information about the current check.
2. Subtract the new check amount from the previous balance.

Write the information you need to solve the problem.

Check information: Northern Electric; check number 7800; date January 10; amount of the check $347.50; previous balance $5,500.00.

Write the system of measurement.

Money

COMPUTE

Do the computation required to complete each task.
Use the formula given above.

previous balance – amount of check = new balance

$5,500.00 – $347.50 = $5,152.50

COMMUNICATE
Enter the information in the checkbook register below.

Check No.	Date	Transaction or Payee	Payment	Deposit	Balance
7799	01/02/9X	Hardik's Office Supply	$650.00		$4,250.00
	01/05/9X	Payment--Olsen's		$1,250.00	$5,500.00
7800	01/10/9X	Northern Electric	$347.50		$5,152.50

PRACTICE

1 You write a check to the Wing Construction Company. The date of the check is January 10. The check number is 7801 and the amount of the check is $786.50. Record the information and the new balance in the check register.

COMPREHEND
In the space below, write what you are to do.

_____

_____

Write the steps you should follow to solve the problem.

_____

_____

Write the information you need to solve the problem.

_____

_____

_____

Write the system of measurement.

_____

COMPUTE
Do the computation required to complete each task.
Use the formula given on page 10.

COMMUNICATE

Enter the information in the checkbook register on page 11.

2 You write a check to Kramer's Air Conditioning on January 12 for $779.37. The check number is 7802. Record the information in the register and find the new balance.

COMPREHEND

In the space below, write what you are to do.

_____

_____

Write the steps you should follow to solve the problem.

_____

_____

Write the information you need to solve the problem.

_____

_____

_____

Write the system of measurement.

_____

COMPUTE

Do the computation required to complete each task.
Use the formula given on page 10.

COMMUNICATE

Enter the information in the checkbook register on page 11.

3 You write a check to the bank First Federal on January 13. The check number is 7803. The amount of the check is $1,010. Record the information and find the new balance.

COMPREHEND

In the space below, write what you are to do.

_____

_____

Write the steps you should follow to solve the problem.

_____

_____

Write the information you need to solve the problem.

_____

_____

_____

Write the system of measurement.

_____

COMPUTE
Do the computation required to complete each task.
Use the formula given on page 10.

COMMUNICATE
Enter the information in the checkbook register on page 11.

TASK 2: CHECKBOOK REGISTER—RECORDING DEPOSITS

Part of the job of keeping a checkbook register is recording deposits to the account. You record each deposit in the checking account register. For each deposit you must:

1. Record all the information about the current check—the date, the amount of deposit, and the name of the person or company that gave you the check or cash to deposit. You don't need a check number because you did not write a check.

2. Find the new checking account balance—the amount of money now in the account. Use this formula:

   previous balance + amount of deposit = new balance

USING THE 3 CS
You make out a bank deposit slip for a payment made to Gibson Paper Company. The check comes from Jackson Stationers. It is for $3,500. The date is January 14. Record the deposit and figure the new balance.

COMPREHEND
In the space below, write what you are to do.

> *Record the deposit in the register and find the new balance.*

Write the steps you should follow to solve the problem.

> *1. Record all the information about the deposit.*
> *2. Add the deposit to the previous balance.*

Write the information you need to solve the problem.

Deposit information: Jackson Stationers; date January 14; amount of the deposit $3,500.00; previous balance $2,576.63 (found in the check register).

Write the system of measurement.

Money

COMPUTE

Do the computation required to complete each task.
Use the formula given on page 13.

previous balance + amount of deposit = new balance

$2,576.63 + $3,500.00 = $6,076.63

COMMUNICATE

Enter the information in the checkbook register below.

Check No.	Date	Transaction or Payee	Payment	Deposit	Balance
					$2,576.63
	01/14/9X	Payment from Jackson Stationers		$3,500.00	$6,076.63

PRACTICE

4 On January 15, you make a deposit of $11,179.02. The check is from Security Envelope. Record the deposit and find the new balance.

COMPREHEND

In the space below, write what you are to do.

_____

Write the steps you should follow to solve the problem.

_____

_____

Write the information you need to solve the problem.

_____

_____

_____

Write the system of measurement.

_____

COMPUTE
Do the computation required to complete each task.
Use the formula given on page 13.

COMMUNICATE
Enter the information in the checkbook register on page 14.

5 On January 16 you make a deposit of $525.25. The check is from Wilson Stationery. You also write check number 7804 to Jane Howard for $1,200. Record the information about the deposit and the check. Find the new balance each time.

COMPREHEND
In the space below, write what you are to do.

_____

_____

_____

Write the steps you should follow to solve the problem.

_____

_____

_____

_____

Write the information you need to solve the problem.

_____

_____

Write the system of measurement.

_____

COMPUTE
Do the computation required to complete each task.
Use the formulas given on page 10 and page 13.

COMMUNICATE

Enter the information in the checkbook register on page 14.

6 On January 17, you write a check for $50.00 to B&B Office Supplies. The check number is 7805. You also make a deposit of $514.75 from Party Tyme Supplies. Record the check and the deposit, and find the new balance each time.

COMPREHEND

In the space below, write what you are to do.

_____

_____

_____

Write the steps you should follow to solve the problem.

_____

_____

_____

Write the information you need to solve the problem.

_____

_____

_____

Write the system of measurement.

_____

COMPUTE

Do the computation required to complete each task.
Use the formulas given on page 10 and page 13.

COMMUNICATE

Enter the information in the checkbook register on page 14.

TASK 3: RECONCILING THE BANK
ACCOUNT (CHECK REGISTER)

As the bookkeeper at Gibson Paper, you need to reconcile the monthly bank statement and the check register. The bank statement shows:

❑ All deposits.
❑ All checks paid.

- ❏ Service charges from the bank.
- ❏ The balance at the beginning and end of the month.

Usually, the bank statement and your check register do *not* agree. The differences could be caused by:

- ❏ Outstanding checks: checks that were recorded in the check register but have not yet been received at the bank for payment.
- ❏ Unrecorded deposits: deposits that were made after the closing date for the bank statement.
- ❏ Bank service fees or credits that you have not yet recorded in the register.

These three things are not errors. But you must find the **adjusted balance** for the account. The adjusted balance takes into account outstanding checks, unrecorded deposits, and bank fees and credits. The adjusted balance for your register should agree with the adjusted balance on the bank statement. If the adjusted balance does not agree, you need to look for an error in the check register. If you cannot find one, you need to check the bank records to see whether the bank made an error.

The first step in reconciling the account is to get an adjusted balance for the check register. To do this:

- ❏ Find the total of the additions to the account. These are credits from the bank to the account.
- ❏ Add these additions to the account balance.
- ❏ Find the total of the deductions to the account. These are bank fees.
- ❏ Subtract the deductions from the account balance to get the new balance.

USING THE 3 CS

You receive the monthly statement for October. You see that the bank has credited you with $5. This is a refund from the bank for a fee from August that was incorrect. The bank is correcting its error. You also see that the bank has charged a service fee of $3. Your check register shows $2,500. Calculate the adjusted check register balance.

COMPREHEND
In the space below, write what you are to do.

Find the adjusted balance for the check register.

Write the steps you should follow to solve the problem.
1. Find the total of the additions to the account.
2. Add the additions to the balance.
3. Find the total of the deductions to the account.
4. Subtract the deductions from the balance.

Write the information you need to solve the problem.

Register balance of $2,500.00; credit of $5.00; fees of $3.00.

Write the system of measurement.

Money

COMPUTE
Do the computation required to complete each task.
Use the steps listed on page 17.

Register balance	*$2,500.00*
Additions	*5.00*
	$2,505.00
Deductions	*3.00*
	$2,502.00

COMMUNICATE
Enter the information in the checkbook register below.

Check No.	Date	Transaction or Payee	Payment	Deposit	Balance
6324	10/31/9X	Bill Wilson	$100.00		$2,500.00
	10/31/9X	Bank Credit		$5.00	$2,505.00
	10/31/9X	Bank Fees (October)	$3.00		$2,502.00

PRACTICE
7 You receive the monthly statement for November. You see that the bank fee is $8.00. Your check register shows $1,830.50. Calculate the adjusted check register balance.

COMPREHEND
In the space below, write what you are to do.

_____

Write the steps you should follow to solve the problem.

_____

_____

_____

Write the information you need to solve the problem.

_____

Write the system of measurement.

_____

COMPUTE
Do the computation required to complete each task.
Use the steps listed on page 17.

COMMUNICATE
Enter the information in the checkbook register below.

Check No.	Date	Transaction or Payee	Payment	Deposit	Balance
6328	11/30/9X	Jackson Electric	$423.00		$1,830.50

8 You receive the monthly statement for December. You see that the bank fee is $7.00. The bank has given you a credit of $7.50 that it deducted incorrectly for a check that it returned (bounced) in error in September. Your check register shows $5,235.27. Calculate the adjusted check register balance.

COMPREHEND
In the space below, write what you are to do.

_____

Write the steps you should follow to solve the problem.

_____

_____

_____

_____

Write the information you need to solve the problem.

_____

Write the system of measurement.

_____

COMPUTE
Do the computation required to complete each task.
Use the steps listed on page 17.

COMMUNICATE
Enter the information in the checkbook register below.

Check No.	Date	Transaction or Payee	Payment	Deposit	Balance
6333	12/28/9X	Federal Express	$221.00		$5,235.27

TASK 4: RECONCILING THE BANK ACCOUNT (BANK STATEMENT)

The second step in reconciling the account is to get an adjusted balance for the statement and then to compare it to the adjusted balance for the check register. To do this:

❑ Find the total of unrecorded deposits. These are deposits listed in the check register but not on the bank statement.

❑ Add the total of deposits to the statement balance.

❑ Find the total of outstanding checks. These are checks recorded in the register but not shown on the statement.

❑ Subtract the total of checks from the statement balance to get the adjusted balance.

❑ Compare the adjusted checkbook register balance to the adjusted bank statement balance. They should be the same.

USING THE 3 CS
You have an adjusted checkbook register balance of $2,502 for October. Here is your October bank statement. Adjust the balance of the bank statement. Compare the two adjusted balances.

Checkbook Register

Check No.	Date	Transaction or Payee	Payment	Deposit	Balance
					$3,015.20
6319	10/05/9X	Jackson Supplies	$720.00		$2,295.20
6320	10/15/9X	Northern Electric	$345.20		$1,950.00
	10/20/9X	Ross Products		$1,000.00	$2,950.00
6323	10/28/9X	Karen Nelson	$550.00		$2,400.00
	10/31/9X	Superior Envelope		$200.00	$2,600.00
6324	10/31/9X	Bill Wilson	$100.00		$2,500.00
	10/31/9X	Bank Credit		$5.00	$2,505.00
	10/31/9X	Bank Fees (October)	$3.00		$2,502.00

Bank Statement

Gibson Paper Company
Account: 30456

Date: October 31, 199X

Beginning Balance: $3,015.20
Amount of Checks/Debits: $1,068.20
Amount of Deposits/Credits: $1,005.00
Ending Balance: $2,952.00

Date	Check No.	Debits	Credits	Balance
10/08/9X	6319	$720.00		$2,295.20
10/18/9X	6320	$345.20		$1,950.00
10/26/9X			$1,000.00	$2,950.00
10/31/9X			$5.00	$2,955.00
10/31/9X	Bank Fees	$3.00		$2,952.00

COMPREHEND

In the space below, write what you are to do.

Find the adjusted balance for the bank statement.

Write the steps you should follow to solve the problem.

1. Find the total of the unrecorded deposits.
2. Add the total deposits to the bank statement balance.
3. Find the total of the outstanding checks.
4. Subtract the total of the outstanding checks.
5. Compare the adjusted balances.

Write the information you need to solve the problem.

Statement balance: $2,952.00
Unrecorded deposits shown in the check register:
$200 (Superior Envelope)
Outstanding checks shown in the check register:
$550 (Karen Nelson) and $100 (Bill Wilson)

Write the system of measurement.

Money

COMPUTE
Do the computation required to complete each task.
Use the steps listed on page 20.

Total unrecorded deposits:	$200	$ 200.00
+ Statement balance:		$2,952.00
Subtotal		$3,152.00
– Total outstanding checks: ($550 + $100)		$ 650.00
Adjusted Balance		$2,502.00

COMMUNICATE
Enter the information in the reconciliation worksheet. Note whether the statement and register agree.

Reconciliation Worksheet	
A. Ending Balance on Bank Statement:	$2,952.00
B. Deposits Made After Last Entry on Statement:	$ 200.00
C. Subtotal (Add A + B.):	$3,152.00
D. Total Outstanding Checks:	$ 650.00
Total (Subtract D from C.)	$2,502.00
(This amount should equal your checkbook balance.)	
Account is reconciled.	

PRACTICE

9 You have an adjusted checkbook register balance of $1,822.50 for November. Here is the November bank statement. Adjust the balance for this bank statement. Compare the two adjusted balances.

Checkbook Register

Check No.	Date	Transaction or Payee	Payment	Deposit	Balance
					$2,502.00
6325	11/05/9X	Tasha Supplies	$650.00		$1,852.00
	11/07/9X	Superior Envelope		$1,000.50	$2,852.50
6326	11/15/9X	B&B Supplies	$700.00		$2,152.50
6327	11/20/9X	Telephone Co.	$699.00		$1,453.50
	11/29/9X	F.J.B. Co.		$800.00	$2,253.50
6328	11/30/9X	Jackson Electric	$423.00		$1,830.50
	11/30/9X	Bank Fees	$ 8.00		$1,822.50

Bank Statement

Gibson Paper Company
Account: 30456

Date: November 30, 199X

Beginning Balance: $2,952.00
Amount of Checks/Debits: $2,707.00
Amount of Deposits/Credits: $1,200.50
Ending Balance: $1,445.50

Date	Check No.	Debits	Credits	Balance
11/01/9X			$200.00	$3,152.00
11/01/9X	6323	$550.00		$2,602.00
11/02/9X	6324	$100.00		$2,502.00
11/05/9X	6325	$650.00		$1,852.00
11/08/9X			$1,000.50	$2,852.50
11/17/9X	6326	$700.00		$2,152.50
11/22/9X	6327	$699.00		$1,453.50
11/30/9X	Bank Fees	$8.00		$1,445.50

COMPREHEND

In the space below, write what you are to do.

_____

_____

Write the steps you should follow to solve the problem.

_____

_____

_____

_____

Write the information you need to solve the problem.

_____

_____

_____

Write the system of measurement.

_____

COMPUTE

Do the computation required to complete each task.
Use the steps listed on page 20.

COMMUNICATE

Enter the information in the reconciliation worksheet. Note whether
the statement and register agree.

Reconciliation Worksheet
A. Ending Balance on Bank Statement:
B. Deposits Made After Last Entry on Statement:
C. Subtotal (Add A + B.):
D. Total Outstanding Checks:
Total (Subtract D from C.)
(This amount should equal your checkbook balance.)
Account is reconciled.

10 You have an adjusted checkbook register balance of $5,235.77 for December. Here is the December bank statement. Adjust the balance for this bank statement. Compare the two adjusted balances.

Checkbook Register

Check No.	Date	Transaction or Payee	Payment	Deposit	Balance
					$1,822.50
6329	12/07/9X	Ross Paper	$500.25		$1,322.25
	12/10/9X	Washington Distributors		$7,500.00	$8,822.25
6330	12/12/9X	J. Burns	$63.98		$8,758.27
6331	12/15/9X	Cold Spring Supply	$3,000.00		$5,758.27
6332	12/20/9X	AAA Auto	$1,500.00		$4,258.27
	12/28/9X	Anderson Stationery		$1,198.00	$5,456.27
6333	12/28/9X	Federal Express	$221.00		$5,235.27
	12/31/9X	Bank Credit		$7.50	$5,242.77
	12/31/9X	Bank Fees	$7.00		$5,235.77

Bank Statement

Gibson Paper Company
Account: 30456

Date: December 31, 199X

Beginning Balance: $1,445.50
Amount of Checks/Debits: $3,994.23
Amount of Deposits/Credits: $8,307.50
Ending Balance: $5,758.77

Date	Check No.	Debits	Credits	Balance
12/01/9X			$800.00	$2,245.50
12/01/9X	6328	$423.00		$1,822.50
12/09/9X	6329	$500.25		$1,322.25
12/12/9X			$7,500.00	$8,822.25
12/14/9X	6330	$63.98		$8,758.27
12/17/9X	6331	$3000.00		$5,758.27
12/31/9X	Bank Credit		$7.50	$5,765.77
12/31/9X	Bank Fees	$7.00		$5,758.77

COMPREHEND

In the space below, write what you are to do.

_____

_____

Write the steps you should follow to solve the problem.

_____

_____

_____

_____

Write the information you need to solve the problem.

_____

_____

_____

_____

Write the system of measurement.

_____

COMPUTE
Do the computation required to complete each task.
Use the steps listed on page 20.

COMMUNICATE
Enter the information in the reconciliation worksheet. Note whether the statement and register agree.

Reconciliation Worksheet

A. Ending Balance on Bank Statement:

B. Deposits Made After Last Entry on Statement:

C. Subtotal (Add A + B.):

D. Total Outstanding Checks:

Total (Subtract D from C.)

(This amount should equal your checkbook balance.)
Account is reconciled.

❏ JOB SITUATION 2 PAYROLL REGISTER

One of the jobs in the Gibson Paper Accounting Department is that of the payroll clerk. The payroll clerk calculates the pay for each company employee.

As payroll clerk, you need several pieces of information to do your job. You need to know the **pay period**, how often the staff is paid. Common pay periods are weekly, biweekly (once every two weeks), twice a month (usually on the fifteenth and thirtieth), and monthly (once a month). A company will not change the pay period that it uses. However, a company may have two different pay periods for two different groups of employees.

You also need to know whether the employee is paid an **hourly wage** or a **salary**. An hourly wage is an amount of money paid for each hour worked. Hourly workers are usually paid an overtime rate for every hour that they work over 40 hours a week. A salary is a fixed amount of money paid for the year. Salaried workers usually do not get paid overtime.

You need to calculate the employee's **gross pay** for the pay period. This is the total amount that the employee has earned. Then you need to calculate **deductions** for taxes. The amount left over after deductions are taken out is the employee's **take-home** or **net pay**.

Here's how you put that information to work as a payroll clerk. You know that hourly workers at Gibson Paper are paid weekly. Each Monday, you receive the records of how much time these employees worked the week before. You must complete the **payroll register** for these employees by Wednesday. A payroll register, such as the one on page 28, shows each employee's gross pay, deductions, and take-home pay.

You use the payroll register to write paychecks like the one shown in Figure 9-6 in the *Knowledge Base*. You make sure the paychecks are signed on Thursday. On Friday morning, you give the paychecks to the employees' supervisors, and they give them to the employees before noon. This means that the employees can deposit their checks in their banks on Friday at lunch and get cash for the weekend. Businesses are never late with paychecks, so you must get your work done on time!

TASK 1: FIGURING GROSS PAY

As the payroll clerk at Gibson Paper Company, you write paychecks for the employees on the payroll register on page 28. These employees are paid an hourly wage for a weekly pay period.

Your first task is to find the gross pay of those employees who worked only **regular hours** this week—40 hours or less. Follow these steps:

1. Find the total number of hours worked.
2. To find their regular pay, use this formula:

$$\text{regular hours} \times \text{regular rate} = \text{regular pay}$$

When there are no overtime hours, regular pay is gross pay. Record it in the payroll register.

USING THE 3 CS
Last week, Mark Bayard worked the following hours:

Monday	8 hours
Tuesday	8 hours
Wednesday	8 hours
Thursday	8 hours
Friday	8 hours

His hourly rate is shown on the payroll register. Calculate his gross pay for the week. Write the total hours, regular pay, and gross pay in the payroll register.

COMPREHEND
In the space below, write what you are to do.

Find Mark Bayard's gross pay. Write the total hours, regular pay, and gross pay in the payroll register.

Write the steps you should follow to solve the problem.

1. Find the total number of hours worked.
2. Find the regular pay (gross pay).

Write the information you need to solve the problem.

Regular Pay Rate: $8.90 per hour.
Hours Worked:

Monday	8 hours
Tuesday	8 hours
Wednesday	8 hours
Thursday	8 hours
Friday	8 hours

Write the system of measurement.

Time and money

COMPUTE
Do the computation required to complete each task.
Use the steps listed on page 27.

1. Total number of hours: $8 + 8 + 8 + 8 + 8 = 40$
2. Regular hours $\times$ regular rate = regular pay
$$40 \times \$8.90 = \$356$$
Regular pay is the gross pay.

COMMUNICATE
Enter the information in the payroll register.

Enter the total hours, regular pay, and gross pay.

PAYROLL REGISTER   Week Ending 9/26

EMPLOYEE INFORMATION   **GROSS EARNINGS**   **DEDUCTIONS**

NAME	MAR STAT	ALLOW	TOTAL HOURS	REG RATE	REG PAY	OT PAY	GROSS PAY	FICA	FWT	NET PAY
M. Bayard	M	4	40	8.90	356.00		356.00			
S. Chester	S	1		8.45						
B. Chu	M	5		6.70						
B. Lopez	S	1		8.45						
S. Marino	M	6		8.40						
C. Masillo	S	3		5.75						
J. Nitti	M	2		13.00						
G. Pitini	M	0		12.45						
J. Stein	M	3		9.60						

PRACTICE
We will skip Sylvia Chester and Ben Chu in the payroll register for now
because they had overtime hours. We'll handle that in Task 2.

1 The time records show that last week, Barbara Lopez worked the following hours:

Monday	8 hours
Tuesday	8 hours
Wednesday	8 hours
Thursday	8 hours
Friday	8 hours

Her hourly rate is shown on the payroll register. Calculate her gross pay for the week. Write the total hours, regular pay, and gross pay in the payroll register.

COMPREHEND

In the space below, write what you are to do.

_____

_____

Write the steps you should follow to solve the problem.

_____

_____

Write the information you need to solve the problem.

_____

_____

_____

_____

_____

_____

_____

Write the system of measurement.

_____

COMPUTE

Do the computation required to complete each task.
Use the steps listed on page 27.

COMMUNICATE

Enter the information in the payroll register on page 28.

2 Last week, Susan Marino worked the following hours:

Monday	8 hours
Tuesday	8 hours
Wednesday	8 hours
Thursday	7 hours
Friday	5 hours

Her hourly rate is shown on the payroll register. Calculate her gross pay for the week. Write the total hours, regular pay, and gross pay in the payroll register.

COMPREHEND
In the space below, write what you are to do.

_____

_____

Write the steps you should follow to solve the problem.

_____

_____

Write the information you need to solve the problem.

_____

_____

_____

_____

_____

_____

_____

Write the system of measurement.

_____

COMPUTE
Do the computation required to complete each task.
Use the steps listed on page 27.

COMMUNICATE
Enter the information in the payroll register on page 28.

3 Last week, Joseph Nitti worked the following hours:

Monday	9 hours
Tuesday	8 hours
Wednesday	7 hours
Thursday	8 hours
Friday	8 hours

His hourly rate is shown on the payroll register. Calculate his gross pay for the week. Write the total hours, regular pay, and gross pay in the payroll register.

COMPREHEND
In the space below, write what you are to do.

_____

_____

Write the steps you should follow to solve the problem.

_____

_____

Write the information you need to solve the problem.

_____

_____

_____

_____

_____

_____

Write the system of measurement.

_____

COMPUTE
Do the computation required to complete each task.
Use the steps listed on page 27.

COMMUNICATE
Enter the information in the payroll register on page 28.

4 Last week, Grace Pitini worked these hours:

Monday	7 hours
Tuesday	7 hours
Wednesday	7 hours
Thursday	7½ hours
Friday	7 hours

Her hourly rate is shown on the payroll register. Calculate her gross pay for the week. Write the total hours, regular pay, and gross pay in the payroll register.

COMPREHEND
In the space below, write what you are to do.

_____

_____

Write the steps you should follow to solve the problem.

_____

_____

Write the information you need to solve the problem.

_____

_____

_____

_____

_____

_____

_____

Write the system of measurement.

_____

COMPUTE
Do the computation required to complete each task.
Use the steps listed on page 27.

COMMUNICATE
Enter the information in the payroll register on page 28.

Part 4: Mathematics and Financial Resources

TASK 2: FIGURING OVERTIME AND GROSS PAY

Some of the employees on the payroll register have worked more than 40 hours in a week. Any hours beyond the regular 40 hours are called **overtime hours**. The rate of pay for these hours is called the **overtime rate**. It is 1½ times the regular rate at Gibson. To figure gross pay for these employees, follow these steps:

1. Find the total number of hours worked. Subtract 40 hours from the total to find the number of overtime hours.
2. Figure the employee's regular pay for the first 40 hours as usual.
3. Figure the employee's overtime rate. Multiply the regular rate by 1½ (or 1.5).
4. Figure the employee's overtime pay. Use this formula:

   overtime hours × overtime rate = overtime pay

5. Find the gross pay. Add the regular pay and overtime pay.

   regular pay + overtime pay = gross pay

USING THE 3 CS

The time records show that Sylvia Chester worked more than 40 hours last week. These are her hours:

Monday	8 hours
Tuesday	9 hours
Wednesday	10 hours
Thursday	8 hours
Friday	8½ hours

Her hourly rate is shown on the payroll register. Calculate her gross pay for the week. Write the total hours, regular pay, overtime pay, and gross pay in the payroll register.

COMPREHEND

In the space below, write what you are to do.

Find the gross pay. Write the total hours, regular pay, overtime pay, and gross pay in the payroll register.

Write the steps you should follow to solve the problem.

1. Find the total number of hours and overtime hours.
2. Figure the regular pay for the first 40 hours.
3. Figure the overtime rate.
4. Figure the overtime pay.
5. Add the regular pay and overtime pay.

Write the information you need to solve the problem.

Monday	*8 hours*
Tuesday	*9 hours*
Wednesday	*10 hours*
Thursday	*8 hours*
Friday	*8½ hours*

Regular rate: $8.45

Write the system of measurement.

Time and money

COMPUTE

Do the computation required to complete each task.
Use the formulas given on page 33.

Total time: $8 + 9 + 10 + 8 + 8\frac{1}{2} = 43\frac{1}{2}$ or 43.5
Overtime hours: $43.5 - 40 = 3\frac{1}{2}$ or 3.5
Regular pay: $40 \times \$8.45 = \$338$
Overtime rate: $\$8.45 \times 1.5 = \$12.68$
Overtime pay: $3.5 \times \$12.68 = \$44.38$
Gross pay: $\$338 + \$44.38 = \$382.38$

COMMUNICATE

Enter the information in the payroll register below.

PAYROLL REGISTER   Week Ending 9/26

EMPLOYEE INFORMATION			GROSS EARNINGS					DEDUCTIONS		
NAME	MAR STAT	ALLOW	TOTAL HOURS	REG RATE	REG PAY	OT PAY	GROSS PAY	FICA	FWT	NET PAY
S. Chester	S	1	43.5	8.45	338.00	44.38	382.38			

PRACTICE

[5] Last week, Ben Chu worked these hours:

Monday	9 hours
Tuesday	9 hours
Wednesday	9 hours
Thursday	9 hours
Friday	9 hours

His hourly rate is shown on the payroll register. Calculate his gross pay for the week. Write the total hours, regular pay, overtime pay, and gross pay in the payroll register.

COMPREHEND

In the space below, write what you are to do.

_____

_____

Write the steps you should follow to solve the problem.

_____

_____

_____

_____

_____

Write the information you need to solve the problem.

_____

_____

_____

_____

_____

Write the system of measurement.

_____

COMPUTE
Do the computation required to complete each task.
Use the formulas given on page 33.

COMMUNICATE
Enter the information in the payroll register on page 28.

6 | Last week, Cindy Masillo worked these hours:

Monday	9 hours
Tuesday	10 hours
Wednesday	10 hours
Thursday	10 hours
Friday	8 hours

Her hourly rate is shown on the payroll register. Calculate her gross pay
for the week. Write the total hours, regular pay, overtime pay, and gross
pay in the payroll register.

COMPREHEND

In the space below, write what you are to do.

_____

_____

Write the steps you should follow to solve the problem.

_____

_____

_____

_____

_____

Write the information you need to solve the problem.

_____

_____

_____

_____

_____

_____

_____

Write the system of measurement.

_____

COMPUTE

Do the computation required to complete each task.
Use the formulas given on page 33.

COMMUNICATE

Enter the information in the payroll register on page 28.

7 Last week, Janet Stein worked these hours:

Monday	7½ hours	Thursday	8 hours
Tuesday	8 hours	Friday	8½ hours
Wednesday	8½ hours		

Her hourly rate is shown on the payroll register. Calculate her gross pay for the week. Write the total hours, regular pay, overtime pay, and gross pay in the payroll register.

COMPREHEND
In the space below, write what you are to do.

_____

_____

Write the steps you should follow to solve the problem.

_____

_____

_____

_____

_____

Write the information you need to solve the problem.

_____

_____

_____

_____

_____

_____

Write the system of measurement.

_____

COMPUTE
Do the computation required to complete each task.
Use the formulas given on page 33.

COMMUNICATE
Enter the information in the payroll register on page 28.

TASK 3: PAYROLL DEDUCTIONS—FICA

All companies are required to subtract certain taxes from an employee's pay before writing a paycheck. These **deductions** include social security tax and federal income tax. They can include state income tax and other deductions such as disability insurance and health insurance.

The first deduction you need to make and record on the payroll register is the deduction for **FICA—Federal Insurance Contributions Act**. This is the full name for what people call "social security tax." It is for the employee's social security account, which provides retirement and disability benefits for the employee. The FICA tax rate changes every year or so. For our practice examples, assume the rate is 7.65 percent.

To find the amount of FICA tax, multiply the employee's gross pay by 7.65 percent and round to hundredths.

$$\textbf{FICA} = \textbf{gross pay} \times \textbf{0.0765}$$

USING THE 3 CS

As payroll clerk, you have figured the gross pay for each employee. You are ready to figure the deductions from the gross pay. You start with Mark Bayard. Find the amount of FICA tax to deduct from his gross pay. Record it in the payroll register on page 39.

COMPREHEND

In the space below, write what you are to do.

Find the FICA tax for Mark Bayard.

Write the step you should follow to solve the problem.

Multiply gross pay by 7.65%.

Write the information you need to solve the problem.

Gross pay from the payroll register is $356.

Write the system of measurement.

Money

COMPUTE

Do the computation required to complete each task.
Use the formula given above.

Multiply gross pay by 7.65%.
First convert 7.65% to a decimal number so that you can multiply.
7.65% = 0.0765
$356.00 × 0.0765 = $27.23

COMMUNICATE

Enter the information in the payroll register on page 39.

PAYROLL REGISTER   Week Ending 9/26

| | EMPLOYEE INFORMATION | | | GROSS EARNINGS | | | | | DEDUCTIONS | | |
NAME	MAR STAT	ALLOW	TOTAL HOURS	REG RATE	REG PAY	OT PAY	GROSS PAY	FICA	FWT	NET PAY
M. Bayard	M	4	40	8.90	356.00		356.00	27.23		
S. Chester	S	1	43.5	8.45	338.00	44.38	382.38			
B. Chu	M	5	45	6.70	268.00	50.25	318.25			
B. Lopez	S	1	40	8.45	338.00		338.00			
S. Marino	M	6	36	8.40	302.40		302.40			
C. Masillo	S	3	47	5.75	230.00	60.41	290.41			
J. Nitti	M	2	40	13.00	520.00		520.00			
G. Pitini	M	0	35.5	12.45	441.98		441.98			
J. Stein	M	3	40.5	9.60	384.00	7.20	391.20			

PRACTICE

8 Compute the FICA tax for Sylvia Chester. Record it in the payroll register.

COMPREHEND

In the space below, write what you are to do.

_____

Write the step you should follow to solve the problem.

_____

Write the information you need to solve the problem.

_____

Write the system of measurement.

_____

COMPUTE

Do the computation required to complete each task.
Use the formula given on page 38.

COMMUNICATE

Enter the information in the payroll register above.

9 Find the FICA tax for Ben Chu. Record it in the register.

COMPREHEND
In the space below, write what you are to do.

_____

Write the step you should follow to solve the problem.

_____

Write the information you need to solve the problem.

_____

Write the system of measurement.

_____

COMPUTE
Do the computation required to complete each task.
Use the formula given on page 38.

COMMUNICATE
Enter the information in the payroll register on page 39.

10 Find the FICA tax for Barbara Lopez. Record it in the payroll register.

COMPREHEND
In the space below, write what you are to do.

_____

Write the step you should follow to solve the problem.

_____

Write the information you need to solve the problem.

_____

Write the system of measurement.

_____

COMPUTE
Do the computation required to complete each task.
Use the formula given on page 38.

COMMUNICATE
Enter the information in the payroll register on page 39.

TASK 4: FEDERAL TAX DEDUCTIONS AND NET PAY

A second deduction that you need to calculate for each employee is **federal withholding tax (FWT)**. This is a part of the employee's pay that is set aside for income tax. The Gibson Accounting Department deducts this money from each employee's paycheck and then sends the money to the federal government. It's important that you calculate the deductions correctly and keep accurate records. At the end of the year, the amount of tax you deducted is written on the employee's W-2 form. The employee puts that figure on his or her income tax form.

As the payroll clerk, you use a government booklet called *Circular E* to figure federal tax deductions. Sample pages from this booklet are shown in Figure 4-1. The booklet has tax tables for married and single people and for weekly pay periods, biweekly pay periods, monthly pay periods, and so on.

You need to know four items from the payroll register:

❑ Payroll period.
❑ Gross pay.
❑ Whether the employee is married or single.
❑ How many "withholding allowances" the employee has claimed.

When they are hired, employees indicate on W-4 forms how many withholding allowances they want. (See Figure 4-2.) Employees claim allowances for themselves and for dependents. They may also claim allowances if they have mortgages, high medical bills, or other expenses that they know will be deductible on their taxes.

You may also have to deduct state tax and other items like health insurance. In these examples, you'll just calculate federal withholding tax.

To figure an employee's federal withholding tax and net pay, follow these steps:

1. Find the row for the employee's gross pay in the married or single tax table in *Circular E*. Find the column for the correct number of allowances for the employee. Look in that column for the amount of tax to be withheld.

2. Subtract the FICA tax and federal withholding tax from the gross pay to get the net or take-home pay.

gross pay − (FICA + FWT) = net pay

SINGLE Persons–WEEKLY Payroll Period
(For Wages Paid After December 1989)

And the wages are–		And the number of withholding allowances claimed is–										
At least	But less than	0	1	2	3	4	5	6	7	8	9	10
		The amount of income tax to be withheld shall be–										
$0	$25	$0	$0	$0	$0	$0	$0	$0	$0	$0	$0	$0
25	30	1	0	0	0	0	0	0	0	0	0	0
30	35	1	0	0	0	0	0	0	0	0	0	0
35	40	2	0	0	0	0	0	0	0	0	0	0
40	45	3	0	0	0	0	0	0	0	0	0	0
45	50	4	0	0	0	0	0	0	0	0	0	0
50	55	4	0	0	0	0	0	0	0	0	0	0
55	60	5	0	0	0	0	0	0	0	0	0	0
60	65	6	0	0	0	0	0	0	0	0	0	0
65	70	7	1	0	0	0	0	0	0	0	0	0
70	75	7	2	0	0	0	0	0	0	0	0	0
75	80	8	2	0	0	0	0	0	0	0	0	0
80	85	9	3	0	0	0	0	0	0	0	0	0
85	90	10	4	0	0	0	0	0	0	0	0	0
90	95	10	5	0	0	0	0	0	0	0	0	0
95	100	11	5	0	0	0	0	0	0	0	0	0
100	105	12	6	0	0	0	0	0	0	0	0	0
105	110	13	7	1	0	0	0	0	0	0	0	0
110	115	13	8	2	0	0	0	0	0	0	0	0
115	120	14	8	2	0	0	0	0	0	0	0	0
120	125	15	9	3	0	0	0	0	0	0	0	0
125	130	16	10	4	0	0	0	0	0	0	0	0
130	135	16	11	5	0	0	0	0	0	0	0	0
135	140	17	11	5	0	0	0	0	0	0	0	0
140	145	18	12	6	0	0	0	0	0	0	0	0
145	150	19	13	7	1	0	0	0	0	0	0	0
150	155	19	14	8	2	0	0	0	0	0	0	0
155	160	20	14	8	2	0	0	0	0	0	0	0
160	165	21	15	9	3	0	0	0	0	0	0	0
165	170	22	16	10	4	0	0	0	0	0	0	0
170	175	22	17	11	5	0	0	0	0	0	0	0
175	180	23	17	11	5	0	0	0	0	0	0	0
180	185	24	18	12	6	0	0	0	0	0	0	0
185	190	25	19	13	7	1	0	0	0	0	0	0
190	195	25	20	14	8	2	0	0	0	0	0	0
195	200	26	20	14	8	3	0	0	0	0	0	0
200	210	27	21	15	10	4	0	0	0	0	0	0
210	220	29	23	17	11	5	0	0	0	0	0	0
220	230	30	24	18	13	7	1	0	0	0	0	0
230	240	32	26	20	14	8	2	0	0	0	0	0
240	250	33	27	21	16	10	4	0	0	0	0	0
250	260	35	29	23	17	11	5	0	0	0	0	0
260	270	36	30	24	19	13	7	1	0	0	0	0
270	280	38	32	26	20	14	8	2	0	0	0	0
280	290	39	33	27	22	16	10	4	0	0	0	0
290	300	41	35	29	23	17	11	5	0	0	0	0
300	310	42	36	30	25	19	13	7	1	0	0	0
310	320	44	38	32	26	20	14	8	2	0	0	0
320	330	45	39	33	28	22	16	10	4	0	0	0
330	340	47	41	35	29	23	17	11	5	0	0	0
340	350	48	42	36	31	25	19	13	7	1	0	0
350	360	50	44	38	32	26	20	14	8	2	0	0
360	370	51	45	39	34	28	22	16	10	4	0	0
370	380	53	47	41	35	29	23	17	11	5	0	0
380	390	54	48	42	37	31	25	19	13	7	1	0
390	400	56	50	44	38	32	26	20	14	8	3	0
400	410	58	51	45	40	34	28	22	16	10	4	0
410	420	61	53	47	41	35	29	23	17	11	6	0
420	430	64	54	48	43	37	31	25	19	13	7	1
430	440	67	56	50	44	38	32	26	20	14	9	3
440	450	70	58	51	46	40	34	28	22	16	10	4
450	460	72	61	53	47	41	35	29	23	17	12	6
460	470	75	64	54	49	43	37	31	25	19	13	7
470	480	78	67	56	50	44	38	32	26	20	15	9
480	490	81	70	59	52	46	40	34	28	22	16	10
490	500	84	72	61	53	47	41	35	29	23	18	12
500	510	86	75	64	55	49	43	37	31	25	19	13
510	520	89	78	67	56	50	44	38	32	26	21	15
520	530	92	81	70	59	52	46	40	34	28	22	16
530	540	95	84	73	62	53	47	41	35	29	24	18

Figure 4–1.

MARRIED Persons–WEEKLY Payroll Period

(For Wages Paid After December 1989)

And the wages are–		And the number of withholding allowances claimed is–										
At least	But less than	0	1	2	3	4	5	6	7	8	9	10
		The amount of income tax to be withheld shall be–										
$0	$70	$0	$0	$0	$0	$0	$0	$0	$0	$0	$0	$0
70	75	1	0	0	0	0	0	0	0	0	0	0
75	80	2	0	0	0	0	0	0	0	0	0	0
80	85	3	0	0	0	0	0	0	0	0	0	0
85	90	3	0	0	0	0	0	0	0	0	0	0
90	95	4	0	0	0	0	0	0	0	0	0	0
95	100	5	0	0	0	0	0	0	0	0	0	0
100	105	6	0	0	0	0	0	0	0	0	0	0
105	110	6	0	0	0	0	0	0	0	0	0	0
110	115	7	1	0	0	0	0	0	0	0	0	0
115	120	8	2	0	0	0	0	0	0	0	0	0
120	125	9	3	0	0	0	0	0	0	0	0	0
125	130	9	3	0	0	0	0	0	0	0	0	0
130	135	10	4	0	0	0	0	0	0	0	0	0
135	140	11	5	0	0	0	0	0	0	0	0	0
140	145	12	6	0	0	0	0	0	0	0	0	0
145	150	12	6	0	0	0	0	0	0	0	0	0
150	155	13	7	1	0	0	0	0	0	0	0	0
155	160	14	8	2	0	0	0	0	0	0	0	0
160	165	15	9	3	0	0	0	0	0	0	0	0
165	170	15	9	3	0	0	0	0	0	0	0	0
170	175	16	10	4	0	0	0	0	0	0	0	0
175	180	17	11	5	0	0	0	0	0	0	0	0
180	185	18	12	6	0	0	0	0	0	0	0	0
185	190	18	12	6	1	0	0	0	0	0	0	0
190	195	19	13	7	1	0	0	0	0	0	0	0
195	200	20	14	8	2	0	0	0	0	0	0	0
200	210	21	15	9	3	0	0	0	0	0	0	0
210	220	22	17	11	5	0	0	0	0	0	0	0
220	230	24	18	12	6	0	0	0	0	0	0	0
230	240	25	20	14	8	2	0	0	0	0	0	0
240	250	27	21	15	9	3	0	0	0	0	0	0
250	260	28	23	17	11	5	0	0	0	0	0	0
260	270	30	24	18	12	6	0	0	0	0	0	0
270	280	31	26	20	14	8	2	0	0	0	0	0
280	290	33	27	21	15	9	3	0	0	0	0	0
290	300	34	29	23	17	11	5	0	0	0	0	0
300	310	36	30	24	18	12	6	0	0	0	0	0
310	320	37	32	26	20	14	8	2	0	0	0	0
320	330	39	33	27	21	15	9	3	0	0	0	0
330	340	40	35	29	23	17	11	5	0	0	0	0
340	350	42	36	30	24	18	12	6	1	0	0	0
350	360	43	38	32	26	20	14	8	2	0	0	0
360	370	45	39	33	27	21	15	9	4	0	0	0
370	380	46	41	35	29	23	17	11	5	0	0	0
380	390	48	42	36	30	24	18	12	7	1	0	0
390	400	49	44	38	32	26	20	14	8	2	0	0
400	410	51	45	39	33	27	21	15	10	4	0	0
410	420	52	47	41	35	29	23	17	11	5	0	0
420	430	54	48	42	36	30	24	18	13	7	1	0
430	440	55	50	44	38	32	26	20	14	8	2	0
440	450	57	51	45	39	33	27	21	16	10	4	0
450	460	58	53	47	41	35	29	23	17	11	5	0
460	470	60	54	48	42	36	30	24	19	13	7	1
470	480	61	56	50	44	38	32	26	20	14	8	2
480	490	63	57	51	45	39	33	27	22	16	10	4
490	500	64	59	53	47	41	35	29	23	17	11	5
500	510	66	60	54	48	42	36	30	25	19	13	7
510	520	67	62	56	50	44	38	32	26	20	14	8
520	530	69	63	57	51	45	39	33	28	22	16	10
530	540	70	65	59	53	47	41	35	29	23	17	11
540	550	72	66	60	54	48	42	36	31	25	19	13
550	560	73	68	62	56	50	44	38	32	26	20	14
560	570	75	69	63	57	51	45	39	34	28	22	16
570	580	76	71	65	59	53	47	41	35	29	23	17
580	590	78	72	66	60	54	48	42	37	31	25	19
590	600	79	74	68	62	56	50	44	38	32	26	20
600	610	81	75	69	63	57	51	45	40	34	28	22
610	620	82	77	71	65	59	53	47	41	35	29	23
620	630	84	78	72	66	60	54	48	43	37	31	25

Figure 4–1 (*con't.*).

19**91** Form W-4

**Department of the Treasury
Internal Revenue Service**

Purpose. Complete Form W-4 so that your employer can withhold the correct amount of Federal income tax from your pay.

Exemption From Withholding. Read line 6 of the certificate below to see if you can claim exempt status. *If exempt, complete line 6; but do not complete lines 4 and 5.* No Federal income tax will be withheld from your pay. Your exemption is good for one year only. It expires February 15, 1992.

Basic Instructions. Employees who are not exempt should complete the Personal Allowances Worksheet. Additional worksheets are provided on page 2 for employees to adjust their withholding allowances based on itemized deductions, adjustments to income, or two-earner/two-job situations. Complete all worksheets that apply to your situation. The worksheets will help you figure the number of withholding allowances you are

entitled to claim. However, you may claim fewer allowances than this.

Head of Household. Generally, you may claim head of household filing status on your tax return only if you are unmarried and pay more than 50% of the costs of keeping up a home for yourself and your dependent(s) or other qualifying individuals.

Nonwage Income. If you have a large amount of nonwage income, such as interest or dividends, you should consider making estimated tax payments using Form 1040-ES. Otherwise, you may find that you owe additional tax at the end of the year.

Two-Earner/Two-Jobs. If you have a working spouse or more than one job, figure the total number of allowances you are entitled to claim on all jobs using worksheets from only one Form

W-4. This total should be divided among all jobs. Your withholding will usually be most accurate when all allowances are claimed on the W-4 filed for the highest paying job and zero allowances are claimed for the others.

Advance Earned Income Credit. If you are eligible for this credit, you can receive it added to your paycheck throughout the year. For details, get Form W-5 from your employer.

Check Your Withholding. After your W-4 takes effect, you can use **Pub. 919,** Is My Withholding Correct for 1991?, to see how the dollar amount you are having withheld compares to your estimated total annual tax. Call 1-800-829-3676 to order this publication. Check your local telephone directory for the IRS assistance number if you need further help.

Personal Allowances Worksheet   For 1991, the value of your personal exemption(s) is reduced if your income is over $100,000 ($150,000 if married filing jointly, $125,000 if head of household, or $75,000 if married filing separately). Get Pub. 919 for details.

A   Enter "1" for **yourself** if no one else can claim you as a dependent . . . . **A** _____

B   Enter "1" if: { 1. You are single and have only one job; or
2. You are married, have only one job, and your spouse does not work; or
3. Your wages from a second job or your spouse's wages (or the total of both) are $1,000 or less. } . . **B** _____

C   Enter "1" for your **spouse.** But, you may choose to enter "0" if you are married and have either a working spouse or more than one job (this may help you avoid having too little tax withheld) . . . . **C** _____

D   Enter number of **dependents** (other than your spouse or yourself) whom you will claim on your tax return . . . . . **D** _____

E   Enter "1" if you will file as **head of household** on your tax return (see conditions under "Head of Household," above) . . **E** _____

F   Enter "1" if you have at least $1,500 of **child or dependent care expenses** for which you plan to claim a credit . . . . **F** _____

G   Add lines A through F and enter total here . . . . ▶ **G** _____

For accuracy, do all worksheets that apply.

- If you plan to **itemize or claim adjustments to income** and want to reduce your withholding, see the Deductions and Adjustments Worksheet on page 2.
- If you are **single** and have **more than one job** and your combined earnings from all jobs exceed $27,000 OR if you are **married** and have a **working spouse or more than one job,** and the combined earnings from all jobs exceed $46,000, see the Two-Earner/Two-Job Worksheet on page 2 if you want to avoid having too little tax withheld.
- If **neither** of the above situations applies, **stop here** and enter the number from line G on line 4 of Form W-4 below.

-------------------- **Cut here and give the certificate to your employer. Keep the top portion for your records.** --------------------

Form **W-4**	**Employee's Withholding Allowance Certificate**	OMB No. 1545-0010
Department of the Treasury Internal Revenue Service	▶ **For Privacy Act and Paperwork Reduction Act Notice, see reverse.**	19**91**

1 Type or print your first name and middle initial | Last name | **2** Your social security number

Home address (number and street or rural route)

City or town, state, and ZIP code

3 Marital status
☐ Single   ☐ Married
☐ Married, but withhold at higher Single rate.
Note: *If married, but legally separated, or spouse is a nonresident alien, check the Single box.*

4 Total number of allowances you are claiming (from line G above or from the Worksheets on back if they apply) . . **4** _____

5 Additional amount, if any, you want deducted from each pay . . . . **5** $ _____

6 I claim exemption from withholding and I certify that I meet **ALL** of the following conditions for exemption:
- Last year I had a right to a refund of **ALL** Federal income tax withheld because I had **NO** tax liability; **AND**
- This year I expect a refund of **ALL** Federal income tax withheld because I expect to have **NO** tax liability; **AND**
- This year if my income exceeds $550 and includes nonwage income, another person cannot claim me as a dependent.

If you meet all of the above conditions, enter the year effective and "EXEMPT" here . . . . . . ▶ **6** 19 _____

7 Are you a full-time student? (**Note:** *Full-time students are not automatically exempt.*) . . . . **7** ☐ Yes ☐ No

Under penalties of perjury, I certify that I am entitled to the number of withholding allowances claimed on this certificate or entitled to claim exempt status.

Employee's signature ▶ _____   Date ▶ _____ , 19 ____

8 Employer's name and address (**Employer:** Complete 8 and 10 **only if sending to IRS**) | **9** Office code (optional) | **10** Employer identification number

Figure 4–2.

USING THE 3 CS

Use the payroll register below to figure the federal withholding tax and take-home pay for Mark Bayard. Record the tax and net pay in the payroll register.

COMPREHEND

In the space below, write what you are to do.

Figure the FWT and take-home pay for Mark Bayard and record them in the payroll register.

Write the steps you should follow to solve the problem.

1. Find the FWT in the tax tables.
2. Subtract the FICA and FWT from the gross pay.

Write the information you need to solve the problem.

Gross pay is $356.00; he is married ("MAR STAT" column); he claims 4 allowances ("ALLOW" column); FICA tax is $27.23.

Write the system of measurement.

Money

COMPUTE

Do the computation required to complete each task.
Use the formula given on page 41.

In the Married Persons Weekly Payroll Period table, the correct row is for $350-$360. The payroll deduction for 4 allowances is $20.

gross pay − (FICA + FWT) = net pay

$356.00 − ($27.23 + $20) = $308.77

COMMUNICATE

Enter the information in the payroll register below.

PAYROLL REGISTER   Week Ending 9/26

| EMPLOYEE INFORMATION | | | | GROSS EARNINGS | | | | | DEDUCTIONS | | |
NAME	MAR STAT	ALLOW	TOTAL HOURS	REG RATE	REG PAY	OT PAY	GROSS PAY	FICA	FWT	NET PAY
M. Bayard	M	4	40	8.90	356.00		356.00	27.23	20.00	308.77
S. Chester	S	1	43.5	8.45	338.00	44.38	382.38	29.25		
B. Chu	M	5	45	6.70	268.00	50.25	318.25	24.35		
B. Lopez	S	1	40	8.45	338.00		338.00	25.86		
S. Marino	M	6	36	8.40	302.40		302.40	23.13		
C. Masillo	S	3	47	5.75	230.00	60.41	290.41	22.22		
J. Nitti	M	2	40	13.00	520.00		520.00	39.78		
G. Pitini	M	0	35.5	12.45	441.98		441.98	33.81		
J. Stein	M	3	40.5	9.60	384.00	7.20	391.20	29.93		

11 Find the federal withholding tax and take-home pay for Sylvia Chester.
Record the tax and net pay in the payroll register.

COMPREHEND
In the space below, write what you are to do.

_____

_____

Write the steps you should follow to solve the problem.

_____

_____

Write the information you need to solve the problem.

_____

_____

Write the system of measurement.

_____

COMPUTE
Do the computation required to complete each task.
Use the formula given on page 41.

COMMUNICATE
Enter the information in the payroll register on page 45.

12 Find the federal withholding tax and take-home pay for Ben Chu. Record
them in the payroll register.

COMPREHEND
In the space below, write what you are to do.

_____

_____

Write the steps you should follow to solve the problem.

_____

_____

Write the information you need to solve the problem.

_____

_____

Write the system of measurement.

_____

COMPUTE
Do the computation required to complete each task.
Use the formula given on page 41.

COMMUNICATE
Enter the information in the payroll register on page 45.

13 Find the federal withholding tax and take-home pay for Barbara Lopez.
Record them in the payroll register.

COMPREHEND
In the space below, write what you are to do.

_____

_____

Write the steps you should follow to solve the problem.

_____

_____

Write the information you need to solve the problem.

_____

_____

Write the system of measurement.

_____

COMPUTE
Do the computation required to complete each task.
Use the formula given on page 41.

COMMUNICATE
Enter the information in the payroll register on page 45.

14 Find the federal withholding tax and take-home pay for Susan Marino.
Record them in the payroll register.

COMPREHEND
In the space below, write what you are to do.

_____

_____

Write the steps you should follow to solve the problem.

_____

_____

Write the information you need to solve the problem.

_____

_____

Write the system of measurement.

_____

COMPUTE
Do the computation required to complete each task.
Use the formula given on page 41.

COMMUNICATE
Enter the information in the payroll register on page 45.

Part 4: Mathematics and Financial Resources

15 Find the federal withholding tax and take-home pay for Cindy Masillo. Record them in the payroll register.

COMPREHEND
In the space below, write what you are to do.

_____

_____

Write the steps you should follow to solve the problem.

_____

_____

Write the information you need to solve the problem.

_____

_____

Write the system of measurement.

_____

COMPUTE
Do the computation required to complete each task.
Use the formula given on page 41.

COMMUNICATE
Enter the information in the payroll register on page 45.

TASK 5: PAYROLL REGISTER FOR SALARIED EMPLOYEES

Remember that some employees are paid a fixed salary for the year. At Gibson, salaried employees have a different pay period. They are paid twice a month, or 24 times a year. Your job as payroll clerk is a little different when it comes to filling out a payroll register for the salaried employees.

Salary doesn't depend on the number of hours worked. It is the same for every pay period. To find the gross pay, divide the year's salary by the number of pay periods. Here's an example:

yearly salary ÷ no. pay periods = gross pay

$24,000 ÷ 24 = $1,000

The payroll register for salaried employees looks like the one shown here. Notice that the gross pay for each employee is filled in. (There are no columns for regular pay or overtime pay.)

PAYROLL REGISTER   Week Ending 9/26

EMPLOYEE INFORMATION			GROSS EARNINGS	DEDUCTIONS		
NAME	MAR STAT	ALLOW	GROSS PAY	FICA	FWT	NET PAY
M. Anderson	M	2	1208.33			
J. Brown	S	0	1625.00			
A. Hernandez	M	4	985.00			
S. Jackson	M	3	1166.66			
B. Mendez	S	1	1333.33			

Because gross pay is the same for every pay period, deductions for the FICA tax and federal withholding tax are the same too. Use the same formula as you've used before to find the FICA tax: Multiply the employee's gross pay by 7.65 percent and round to hundredths.

Use the tax tables for semimonthly pay (married or single persons) shown in Figure 4-3 to figure federal withholding tax.

To find net pay, use this formula:

$$\text{gross pay} - (\text{FICA} + \text{FWT}) = \text{net pay}$$

As payroll clerk, you only need to calculate the net pay for salaried employees once a year or when something changes. An employee may change his or her marital status or number of allowances. The government may change the tax rate. Usually, however, your job as payroll clerk is just to make out the checks for the same amount twice a month. Then you get them signed and sent to the employee's supervisor usually on the fifteenth and thirtieth of the month.

USING THE 3 CS

It's the start of the year and you need to calculate taxes and net pay for salaried employees. Find the FICA tax, federal withholding tax, and net pay for Mary Anderson. Record them in the payroll register.

COMPREHEND

In the space below, write what you are to do.

Find the FICA tax, federal withholding tax, and net pay for Mary Anderson. Record them in the payroll register.

Write the steps you should follow to solve the problem.

1. Find the FICA tax by multiplying gross pay by 7.65%.
2. Find the FWT in the tax tables.
3. Find the net pay.

SINGLE Persons–SEMIMONTHLY Payroll Period
(For Wages Paid After December 1989)

And the wages are–		And the number of withholding allowances claimed is–										
At least	But less than	0	1	2	3	4	5	6	7	8	9	10
		The amount of income tax to be withheld shall be–										
$600	$620	$84	$71	$58	$46	$33	$20	$7	$0	$0	$0	$0
620	640	87	74	61	49	36	23	10	0	0	0	0
640	660	90	77	64	52	39	26	13	0	0	0	0
660	680	93	80	67	55	42	29	16	3	0	0	0
680	700	96	83	70	58	45	32	19	6	0	0	0
700	720	99	86	73	61	48	35	22	9	0	0	0
720	740	102	89	76	64	51	38	25	12	0	0	0
740	760	105	92	79	67	54	41	28	15	3	0	0
760	780	108	95	82	70	57	44	31	18	6	0	0
780	800	111	98	85	73	60	47	34	21	9	0	0
800	820	114	101	88	76	63	50	37	24	12	0	0
820	840	117	104	91	79	66	53	40	27	15	2	0
840	860	120	107	94	82	69	56	43	30	18	5	0
860	880	124	110	97	85	72	59	46	33	21	8	0
880	900	130	113	100	88	75	62	49	36	24	11	0
900	920	135	116	103	91	78	65	52	39	27	14	1
920	940	141	119	106	94	81	68	55	42	30	17	4
940	960	147	123	109	97	84	71	58	45	33	20	7
960	980	152	128	112	100	87	74	61	48	36	23	10
980	1,000	158	134	115	103	90	77	64	51	39	26	13
1,000	1,020	163	140	118	106	93	80	67	54	42	29	16
1,020	1,040	169	145	121	109	96	83	70	57	45	32	19
1,040	1,060	175	151	127	112	99	86	73	60	48	35	22
1,060	1,080	180	156	132	115	102	89	76	63	51	38	25
1,080	1,100	186	162	138	118	105	92	79	66	54	41	28
1,100	1,120	191	168	144	121	108	95	82	69	57	44	31
1,120	1,140	197	173	149	125	111	98	85	72	60	47	34
1,140	1,160	203	179	155	131	114	101	88	75	63	50	37
1,160	1,180	208	184	160	136	117	104	91	78	66	53	40
1,180	1,200	214	190	166	142	120	107	94	81	69	56	43
1,200	1,220	219	196	172	148	124	110	97	84	72	59	46
1,220	1,240	225	201	177	153	129	113	100	87	75	62	49
1,240	1,260	231	207	183	159	135	116	103	90	78	65	52
1,260	1,280	236	212	188	164	141	119	106	93	81	68	55
1,280	1,300	242	218	194	170	146	122	109	96	84	71	58
1,300	1,320	247	224	200	176	152	128	112	99	87	74	61
1,320	1,340	253	229	205	181	157	133	115	102	90	77	64
1,340	1,360	259	235	211	187	163	139	118	105	93	80	67
1,360	1,380	264	240	216	192	169	145	121	108	96	83	70
1,380	1,400	270	246	222	198	174	150	126	111	99	86	73
1,400	1,420	275	252	228	204	180	156	132	114	102	89	76
1,420	1,440	281	257	233	209	185	161	138	117	105	92	79
1,440	1,460	287	263	239	215	191	167	143	120	108	95	82
1,460	1,480	292	268	244	220	197	173	149	125	111	98	85
1,480	1,500	298	274	250	226	202	178	154	130	114	101	88
1,500	1,520	303	280	256	232	208	184	160	136	117	104	91
1,520	1,540	309	285	261	237	213	189	166	142	120	107	94
1,540	1,560	315	291	267	243	219	195	171	147	123	110	97
1,560	1,580	320	296	272	248	225	201	177	153	129	113	100
1,580	1,600	326	302	278	254	230	206	182	158	135	116	103
1,600	1,620	331	308	284	260	236	212	188	164	140	119	106
1,620	1,640	337	313	289	265	241	217	194	170	146	122	109
1,640	1,660	343	319	295	271	247	223	199	175	151	127	112
1,660	1,680	348	324	300	276	253	229	205	181	157	133	115
1,680	1,700	354	330	306	282	258	234	210	186	163	139	118
1,700	1,720	359	336	312	288	264	240	216	192	168	144	121
1,720	1,740	365	341	317	293	269	245	222	198	174	150	126
1,740	1,760	371	347	323	299	275	251	227	203	179	155	131
1,760	1,780	376	352	328	304	281	257	233	209	185	161	137
1,780	1,800	382	358	334	310	286	262	238	214	191	167	143
1,800	1,820	387	364	340	316	292	268	244	220	196	172	148
1,820	1,840	393	369	345	321	297	273	250	226	202	178	154
1,840	1,860	399	375	351	327	303	279	255	231	207	183	159
1,860	1,880	404	380	356	332	309	285	261	237	213	189	165
1,880	1,900	410	386	362	338	314	290	266	242	219	195	171

$1,900 and over   Use Table 3(a) for a **SINGLE person** on page 22. Also see the instructions on page 20.

Figure 4–3.

MARRIED Persons–SEMIMONTHLY Payroll Period
(For Wages Paid After December 1989)

And the wages are–		And the number of withholding allowances claimed is–										
At least	But less than	0	1	2	3	4	5	6	7	8	9	10
		The amount of income tax to be withheld shall be–										
$960	$980	$124	$111	$99	$86	$73	$60	$47	$35	$22	$9	$0
980	1,000	127	114	102	89	76	63	50	38	25	12	0
1,000	1,020	130	117	105	92	79	66	53	41	28	15	2
1,020	1,040	133	120	108	95	82	69	56	44	31	18	5
1,040	1,060	136	123	111	98	85	72	59	47	34	21	8
1,060	1,080	139	126	114	101	88	75	62	50	37	24	11
1,080	1,100	142	129	117	104	91	78	65	53	40	27	14
1,100	1,120	145	132	120	107	94	81	68	56	43	30	17
1,120	1,140	148	135	123	110	97	84	71	59	46	33	20
1,140	1,160	151	138	126	113	100	87	74	62	49	36	23
1,160	1,180	154	141	129	116	103	90	77	65	52	39	26
1,180	1,200	157	144	132	119	106	93	80	68	55	42	29
1,200	1,220	160	147	135	122	109	96	83	71	58	45	32
1,220	1,240	163	150	138	125	112	99	86	74	61	48	35
1,240	1,260	166	153	141	128	115	102	89	77	64	51	38
1,260	1,280	169	156	144	131	118	105	92	80	67	54	41
1,280	1,300	172	159	147	134	121	108	95	83	70	57	44
1,300	1,320	175	162	150	137	124	111	98	86	73	60	47
1,320	1,340	178	165	153	140	127	114	101	89	76	63	50
1,340	1,360	181	168	156	143	130	117	104	92	79	66	53
1,360	1,380	184	171	159	146	133	120	107	95	82	69	56
1,380	1,400	187	174	162	149	136	123	110	98	85	72	59
1,400	1,420	190	177	165	152	139	126	113	101	88	75	62
1,420	1,440	193	180	168	155	142	129	116	104	91	78	65
1,440	1,460	196	183	171	158	145	132	119	107	94	81	68
1,460	1,480	199	186	174	161	148	135	122	110	97	84	71
1,480	1,500	202	189	177	164	151	138	125	113	100	87	74
1,500	1,520	207	192	180	167	154	141	128	116	103	90	77
1,520	1,540	213	195	183	170	157	144	131	119	106	93	80
1,540	1,560	219	198	186	173	160	147	134	122	109	96	83
1,560	1,580	224	201	189	176	163	150	137	125	112	99	86
1,580	1,600	230	206	192	179	166	153	140	128	115	102	89
1,600	1,620	235	211	195	182	169	156	143	131	118	105	92
1,620	1,640	241	217	198	185	172	159	146	134	121	108	95
1,640	1,660	247	223	201	188	175	162	149	137	124	111	98
1,660	1,680	252	228	204	191	178	165	152	140	127	114	101
1,680	1,700	258	234	210	194	181	168	155	143	130	117	104
1,700	1,720	263	239	216	197	184	171	158	146	133	120	107
1,720	1,740	269	245	221	200	187	174	161	149	136	123	110
1,740	1,760	275	251	227	203	190	177	164	152	139	126	113
1,760	1,780	280	256	232	208	193	180	167	155	142	129	116
1,780	1,800	286	262	238	214	196	183	170	158	145	132	119
1,800	1,820	291	267	244	220	199	186	173	161	148	135	122
1,820	1,840	297	273	249	225	202	189	176	164	151	138	125
1,840	1,860	303	279	255	231	207	192	179	167	154	141	128
1,860	1,880	308	284	260	236	212	195	182	170	157	144	131
1,880	1,900	314	290	266	242	218	198	185	173	160	147	134
1,900	1,920	319	295	272	248	224	201	188	176	163	150	137
1,920	1,940	325	301	277	253	229	205	191	179	166	153	140
1,940	1,960	331	307	283	259	235	211	194	182	169	156	143
1,960	1,980	336	312	288	264	240	217	197	185	172	159	146
1,980	2,000	342	318	294	270	246	222	200	188	175	162	149
2,000	2,020	347	323	300	276	252	228	204	191	178	165	152
2,020	2,040	353	329	305	281	257	233	209	194	181	168	155
2,040	2,060	359	335	311	287	263	239	215	197	184	171	158
2,060	2,080	364	340	316	292	268	245	221	200	187	174	161
2,080	2,100	370	346	322	298	274	250	226	203	190	177	164
2,100	2,120	375	351	328	304	280	256	232	208	193	180	167
2,120	2,140	381	357	333	309	285	261	237	214	196	183	170
2,140	2,160	387	363	339	315	291	267	243	219	199	186	173
2,160	2,180	392	368	344	320	296	273	249	225	202	189	176
2,180	2,200	398	374	350	326	302	278	254	230	206	192	179
2,200	2,220	403	379	356	332	308	284	260	236	212	195	182
2,220	2,240	409	385	361	337	313	289	265	242	218	198	185
2,240	2,260	415	391	367	343	319	295	271	247	223	201	188

$2,260 and over   Use Table 3(b) for a **MARRIED person** on page 22. Also see the instructions on page 20.

Figure 4–3. (*con't.*)

Write the information you need to solve the problem.

Gross pay = $1,208.33; married; 2 allowances.

Write the system of measurement.

Money

COMPUTE
Do the computation required to complete each task.

FICA: $1,208.33 × 7.65% = $92.44
FWT: $135
Net Pay: $1,208.33 − ($92.44 + $135) = $980.89

COMMUNICATE
Enter the information in the payroll register below.

EMPLOYEE INFORMATION			GROSS EARNINGS	DEDUCTIONS		
NAME	MAR STAT	ALLOW	GROSS PAY	FICA	FWT	NET PAY
M. Anderson	M	2	1208.33	92.44	135.00	980.89

PRACTICE

16 Find the FICA tax, federal withholding tax, and net pay for Jack Brown. Record them in the payroll register.

COMPREHEND
In the space below, write what you are to do.

_____

_____

Write the steps you should follow to solve the problem.

_____

_____

_____

Write the information you need to solve the problem.

_____

Write the system of measurement.

_____

COMPUTE
Do the computation required to complete each task.
Use the formula given on page 50.

COMMUNICATE
Enter the information in the payroll register on page 50.

17 Find the FICA tax, federal withholding tax, and net pay for Anna Hernandez. Record them in the payroll register.

COMPREHEND
In the space below, write what you are to do.

_____

_____

Write the steps you should follow to solve the problem.

_____

_____

_____

Write the information you need to solve the problem.

_____

Write the system of measurement.

_____

COMPUTE
Do the computation required to complete each task.
Use the formula given on page 50.

COMMUNICATE
Enter the information in the payroll register on page 50.

18 Find the FICA tax, federal withholding tax, and net pay for Sara Jackson. Record them in the payroll register.

COMPREHEND
In the space below, write what you are to do.

_____

_____

Write the steps you should follow to solve the problem.

_____

_____

_____

Write the information you need to solve the problem.

_____

Write the system of measurement.

_____

COMPUTE
Do the computation required to complete each task.
Use the formula given on page 50.

COMMUNICATE
Enter the information in the payroll register on page 50.

19 Find the FICA tax, federal withholding tax, and net pay for Barbara Mendez. Record them in the payroll register.

COMPREHEND
In the space below, write what you are to do.

_____

_____

Write the steps you should follow to solve the problem.

_____

_____

_____

Write the information you need to solve the problem.

_____

Write the system of measurement.

_____

COMPUTE
Do the computation required to complete each task.
Use the formula given on page 50.

COMMUNICATE
Enter the information in the payroll register on page 50.

MATHEMATICS AND MATERIAL RESOURCES

PART 5

Part 5 covers some of the material resources you may deal with on the job. There are three job situations in this part of the book: shipping, the Copy Center, and travel arrangements. Each job situation requires you to do several math tasks. Use the 3 Cs to complete each math task.

❏ JOB SITUATION 1   SHIPPING

Every business communication is important. Whether it is a shipment to a customer or a letter with a check to pay a bill, the communication must get through. Put yourself in the role of mail clerk at Gibson Paper. It is your job to make sure that the right postage is put on the letter or package so that the United States Postal Service (U.S.P.S) will deliver it. Chapter 3 of the *Knowledge Base* discusses the different types or **classes** of mail.

❏ As a mail clerk, you use **first-class mail** for ordinary letters. These letters will arrive within a week.

❏ You use **Priority Mail** for larger letters and for packages that need to arrive as quickly as first-class mail.

❏ You use **fourth-class mail**, or **Parcel Post**, for packages that don't need to travel as quickly.

At Gibson, you will also consider using *United Parcel Service* (UPS). UPS is a private company that delivers everywhere that the U.S.P.S. does, except to post office boxes. You always compare to see whether UPS or fourth-class mail is cheaper.

Sometimes it seems that everything that is shipped is a "Rush" item. The quickest way to send something is with an **overnight mail service.** These services, such as the U.S.P.S., Federal Express, and Airborne, can deliver by 10:30 the next business morning. Or they can deliver later. The faster something is delivered, the more it costs to ship it. Your company wants to keep its shipping costs down. So your job is to balance speed with cost when using overnight mail.

TASK 1: FINDING FIRST-CLASS AND PRIORITY MAIL RATES

One of your jobs is to put the correct postage on letters. You use either first-class mail or Priority Mail for these letters. First-class mail is for letters up to 11 ounces. Figure 3-3 in the *Knowledge Base* shows the postal rate table for different weights of letters sent first class.

For letters over 11 ounces, use the Priority Mail rate table in Figure 3-4. The prices are given for different zones. A zone is an area a certain distance from where the package was mailed. For packages up to 5 pounds, the price for each zone is the same. Above 5 pounds, packages going to zones farther away cost more. (You will not deal with zones in Task 1.)

To find the first-class rate for a letter or package, first weigh the letter. Then follow these steps:

❏ Decide whether to use first-class or Priority Mail. Use first class up to 11 ounces. Use Priority Mail over 11 ounces.

❏ Look up the weight in the postal rates table. Notice that each price is for a weight up to *but not over* the weight shown. (The rate for a letter weighing 2 ounces is $0.52. The rate for a letter weighing 2½ ounces is $0.75. That is the rate for a piece of mail up to but not over 3 ounces.)

Note: When using Priority Mail, use the 1-pound rate for any letter or package from 11 ounces to 1 pound.

USING THE 3 CS

You have three letters to send. You weigh them and find they weigh 1 ounce, 13 ounces, and 4¼ ounces. Should each envelope be sent first-class mail or Priority Mail? How much postage does each need?

COMPREHEND

In the space below, write what you are to do.

Find out whether to send the letter first-class mail or Priority Mail. Find how much postage is needed for each letter.

Write the steps you should follow to solve the problem.

1. Decide whether to use first-class mail or Priority Mail.
2. Look up the rate in the postal rate tables in the Knowledge Base.

Write the information you need to solve the problem.

The letters weigh 1 ounce, 13 ounces, and 4¼ ounces.

Write the system of measurement.

Weight and money

COMPUTE

Do the computation required to complete each task.

1. Send First Class: 1-ounce rate = $0.29
2. Send Priority Mail: 13 ounces = 1-pound rate = $2.90
3. Send First Class: 4¼ ounces = 5-ounce rate = $1.21

COMMUNICATE

Write the rate and the type of mail as you would write it on each envelope.

> *$0.29 First Class*
> *$2.90 Priority Mail*
> *$1.21 First Class*

PRACTICE

1   You have three letters to send. You weigh them and find they weigh 2½ ounces, 3 ounces, and 8 ounces. Should each envelope be sent first-class mail or Priority Mail? How much postage does each need?

COMPREHEND

In the space below, write what you are to do.

_____

_____

Write the steps you should follow to solve the problem.

_____

_____

Write the information you need to solve the problem.

_____

Write the system of measurement.

_____

COMPUTE

Do the computation required to complete each task.

_____

_____

_____

COMMUNICATE

Write the rate and type of mail as you would write them on each envelope.

_____

_____

_____

2 You have three letters to send. You weigh them and find they weigh 5 ounces, 1 pound, and 2 ounces. Should each envelope be sent first-class mail or Priority Mail? How much postage does each need?

COMPREHEND
In the space below, write what you are to do.

_____

_____

Write the steps you should follow to solve the problem.

_____

_____

Write the information you need to solve the problem.

_____

Write the system of measurement.

_____

COMPUTE
Do the computation required to complete each task.

_____

_____

_____

COMMUNICATE
Write the rate and type of mail as you would write them on each envelope.

_____

_____

_____

3 You have four letters to send. You weigh them and find they weigh 3½ ounces, 1¼ pounds, 1 ounce, and 4 ounces. Should each envelope be sent first-class mail or Priority Mail? How much postage does each need?

COMPREHEND
In the space below, write what you are to do.

_____

_____

Write the steps you should follow to solve the problem.

_____

_____

Write the information you need to solve the problem.

_____

_____

Write the system of measurement.

_____

COMPUTE
Do the computation required to complete each task.

_____

_____

_____

_____

COMMUNICATE
Write the rate and type of mail as you would write them on each envelope.

_____

_____

_____

_____

4 You have four envelopes to send. They weigh 10½ ounces, 3½ pounds, 7¼ ounces, and just a shade over 11 ounces. Should each envelope be sent first-class mail or Priority Mail? How much postage does each need?

COMPREHEND
In the space below, write what you are to do.

_____

_____

Write the steps you should follow to solve the problem.

_____

_____

Write the information you need to solve the problem.

_____

_____

Write the system of measurement.

_____

COMPUTE

Do the computation required to complete each task.

_____

_____

_____

_____

COMMUNICATE

Write the rate and type of mail as you would write them on each envelope.

_____

_____

_____

_____

5 You have four envelopes to send. They weigh under 1 ounce, 6½ ounces, 5 pounds, and 17 ounces. Should each envelope be sent first-class mail or Priority Mail? How much postage does each need?

COMPREHEND

In the space below, write what you are to do.

_____

_____

Write the steps you should follow to solve the problem.

_____

_____

Write the information you need to solve the problem.

_____

_____

Write the system of measurement.

_____

COMPUTE

Do the computation required to complete each task.

_____

_____

_____

_____

COMMUNICATE

Write the rate and type of mail as you would write them on each envelope.

_____

_____

_____

_____

TASK 2: FINDING FOURTH-CLASS MAIL RATES

As a mail clerk, you also need to put the correct postage on packages that will be mailed. Use fourth-class mail for packages. Figure 3-4 in the _Knowledge Base_ includes the postal rate table for different weights of packages sent fourth class. Prices are given for different delivery zones. Zones are areas within a certain distance from where the package was mailed. For example, Zone 1 is for packages delivered within 50 miles. Zone 2 is farther away. It includes destinations 51 to 100 miles away.

To find the fourth-class rate, first weigh the package. You must also know the zone the package is going to. (You find the zone by looking up the ZIP Code of the address in a table. You will not look up the zone in this Task.)

Once you know the weight and zone, find the rate by this method:

❑ Find the weight in the postal rate table in the _Knowledge Base_. Round up to the next pound if the package weight is over a weight shown on the chart.

❑ Find the rate for the zone to which the package is going.

USING THE 3 CS

You have a package weighing 16 pounds 4 ounces. It is going to Zone 6. How much postage do you need?

COMPREHEND

In the space below, write what you are to do.

Find how much postage is needed.

Write the step you should follow to solve the problem.

Find the rate for the weight and zone.

Write the information you need to solve the problem.

Weight = 16 pounds 4 ounces
Zone = 6

Write the system of measurement.

Weight and money

COMPUTE
Do the computation required to complete each task.

> *Look in Zone 6.*
> *16 pounds 4 ounces = rate for 17 pounds = $9.62*

COMMUNICATE
Write the rate and type of mail as you would write them on the package.
> *$9.62 Parcel Post, or Fourth Class*

PRACTICE

6 You have a package weighing 41 pounds. It is going to Zone 4. How much postage do you need?

COMPREHEND
In the space below, write what you are to do.

_____

Write the step you should follow to solve the problem.

_____

Write the information you need to solve the problem.

_____

_____

Write the system of measurement.

_____

COMPUTE
Do the computation required to complete each task.

_____

_____

COMMUNICATE
Write the rate and type of mail as you would write them on the package.

_____

7 You have a package weighing 8 pounds 12 ounces. It is going to Zone 5. How much postage do you need?

COMPREHEND
In the space below, write what you are to do.

_____

Write the step you should follow to solve the problem.

_____

Write the information you need to solve the problem.

_____

_____

Write the system of measurement.

_____

COMPUTE
Do the computation required to complete each task.

_____

_____

COMMUNICATE
Write the rate and type of mail as you would write them on the package.

_____

8 You have a package weighing 4 pounds 2 ounces. It is going to Zone 1. How much postage do you need?

COMPREHEND
In the space below, write what you are to do.

_____

Write the step you should follow to solve the problem.

_____

Write the information you need to solve the problem.

_____

Write the system of measurement.

_____

COMPUTE
Do the computation required to complete each task.

_____

_____

COMMUNICATE
Write the rate and type of mail as you would write them on the package.

_____

9 You have a package weighing 50 pounds 8 ounces. It is going to Zone 8. How much postage do you need?

COMPREHEND
In the space below, write what you are to do.

_____

Write the step you should follow to solve the problem.

_____

Write the information you need to solve the problem.

_____

_____

Write the system of measurement.

_____

COMPUTE
Do the computation required to complete each task.

_____

_____

COMMUNICATE
Write the rate and type of mail as you would write them on the package.

_____

TASK 3: FINDING UNITED PARCEL SERVICE RATES

Gibson Paper Company also uses United Parcel Service (UPS) to deliver some packages. UPS delivers packages to zones just as the U.S.P.S. does. The zones are almost the same. To find the rate for a UPS package, you use three tables: the Zone Chart shown here in Figure 5-1; the Commercial Deliveries Rate Chart shown in Figure 5-2; and the Residential Deliveries Rate Chart shown in Figure 5-3.

To find the rate for a package, follow these steps:

1. Find the zone in the Zone Chart. Look up the first three digits of the ZIP Code.
2. For packages going to businesses, look up the weight in the Commercial Deliveries Rate Chart. (Round up to the next pound if the package weight is over a weight shown on the chart.) Find the rate for the correct zone.
3. For packages going to homes, look up the weight and zone in the Residential Deliveries Rate Chart.

GROUND SERVICE
ZONE CHART
For Shippers with ZIP Codes 068-01 to 069-99

Service to 48 Continental United States
To determine zone, take first three digits of ZIP Code to which parcel is addressed and refer to chart below.

ZIP CODE PREFIXES	UPS ZONE	ZIP CODE PREFIXES	UPS ZONE	ZIP CODE PREFIXES	UPS ZONE	ZIP CODE PREFIXES	UPS ZONE
004-005	2	220-227	3	430-459	4	733	7
010-018	2	228-229	4	460-466	5	734-738	6
019	3	230-232	3	467-468	4	739	7
020-024	2	233-253	4	469	5	740-762	6
025-026	3	254	3	470	4	763-772	7
027-034	2	255-266	4	471-472	5	773	6
035	3	267	3	473	4	774-775	7
036	2	268-288	4	474-479	5	776-777	6
037-043	3	289-294	5	480-489	4	778-797	7
044	4	295	4	490-491	5	798-799	8
045	3	296	5	492	4		
046-047	4	297	4	493-499	5	800-812	7
048-050	3	298-299	5			813	8
051-053	2			500-503	6	814	7
054	3	300-324	5	504	5	815	8
055	2	325	6	505	6	816-820	7
056-059	3	326-329	5	506-507	5	821	8
060-089	2	330-337	6	508-516	6	822-828	7
		338	5	520-559	5	829-874	8
100-128	2	339	6	560-576	6	875-877	7
129-132	3	342-346	6	577	7	878-880	8
133-135	2	347	5	580-585	6	881-882	7
136	3	349	6	586-593	7	883	8
137-139	2	350-364	5	594	8	884	7
140-149	3	365-366	6	595	7	885-899	8
150-154	4	367-368	5	596-599	8		
155	3	369	6			900-961	8
156	4	370-375	5	600-639	5	970-986	8
157-160	3	376	4	640-649	6	988-994	8
161	4	377-386	5	650-652	5		
162-163	3	387	6	653	6		
164-165	4	388-389	5	654-655	5		
166-174	3	390-396	6	656-676	6		
175-176	2	397-399	5	677-679	7		
177	3			680-692	6		
178-198	2	400-402	5	693	7		
199	3	403-405	4				
		406-409	5	700-722	6		
200-218	3	410-418	4	723-724	5		
219	2	420-427	5	725-732	6		

See separate charts for UPS Next Day Air, 2nd Day Air and International services.
Air service is provided to all points in Alaska, Hawaii and Puerto Rico.

017232 REV. 2-89

Figure 5-1.

Commercial Deliveries
Delivery to a place of business†

WEIGHT NOT TO EXCEED	2	3	4	5	6	7	8
1 lb.	$ 1.88	$ 2.01	$ 2.18	$ 2.25	$ 2.32	$ 2.41	$ 2.49
2 "	1.89	2.02	2.42	2.51	2.68	2.80	3.00
3 "	1.97	2.18	2.60	2.73	2.96	3.18	3.44
4 "	2.07	2.34	2.73	2.91	3.22	3.45	3.79
5 "	2.16	2.43	2.78	2.97	3.36	3.61	4.01
6 "	2.26	2.48	2.81	2.99	3.39	3.77	4.23
7 "	2.33	2.51	2.84	3.02	3.47	3.97	4.47
8 "	2.42	2.55	2.87	3.21	3.74	4.29	4.87
9 "	2.52	2.64	2.96	3.39	4.00	4.61	5.27
10 "	2.60	2.71	3.09	3.56	4.23	4.93	5.64
11 "	2.68	2.80	3.23	3.78	4.48	5.27	6.08
12 "	2.77	2.92	3.40	3.99	4.79	5.62	6.50
13 "	2.84	3.00	3.58	4.21	5.08	5.99	6.93
14 "	2.88	3.13	3.77	4.42	5.36	6.33	7.36
15 "	2.95	3.26	3.93	4.65	5.64	6.70	7.79
16 "	3.03	3.39	4.10	4.88	5.94	7.05	8.22
17 "	3.08	3.52	4.27	5.09	6.22	7.40	8.65
18 "	3.17	3.64	4.44	5.31	6.50	7.76	9.07
19 "	3.29	3.80	4.62	5.52	6.80	8.12	9.50
20 "	3.41	3.93	4.79	5.75	7.08	8.46	9.93
21 "	3.54	4.06	4.97	5.98	7.36	8.83	10.36
22 "	3.64	4.19	5.14	6.19	7.65	9.18	10.79
23 "	3.75	4.31	5.31	6.41	7.94	9.53	11.21
24 "	3.85	4.44	5.48	6.63	8.23	9.89	11.65
25 "	3.95	4.56	5.65	6.85	8.50	10.25	12.07
26 "	4.05	4.69	5.83	7.08	8.80	10.59	12.50
27 "	4.15	4.84	6.01	7.29	9.08	10.96	12.94
28 "	4.25	4.96	6.18	7.51	9.37	11.31	13.35
29 "	4.34	5.09	6.35	7.74	9.66	11.67	13.80
30 "	4.44	5.22	6.54	7.98	9.98	12.06	14.27
31 "	4.53	5.39	6.75	8.24	10.30	12.44	14.74
32 "	4.62	5.55	6.94	8.50	10.63	12.84	15.21
33 "	4.71	5.69	7.15	8.75	10.95	13.23	15.68
34 "	4.79	5.84	7.33	8.96	11.25	13.61	16.15
35 "	4.89	5.96	7.49	9.19	11.54	14.01	16.59
36 "	4.97	6.11	7.67	9.42	11.83	14.35	17.04
37 "	5.06	6.25	7.85	9.64	12.12	14.73	17.47
38 "	5.15	6.38	8.02	9.87	12.42	15.08	17.91
39 "	5.23	6.51	8.21	10.08	12.71	15.44	18.34
40 "	5.32	6.63	8.37	10.31	12.99	15.81	18.77

Figure 5–2.

Residential Deliveries
Delivery to a home†

WEIGHT NOT TO EXCEED	2	3	4	5	6	7	8
1 lb.	$ 2.18	$ 2.31	$ 2.48	$ 2.55	$ 2.62	$ 2.71	$ 2.79
2 "	2.19	2.32	2.72	2.81	2.98	3.10	3.30
3 "	2.27	2.48	2.90	3.03	3.26	3.48	3.74
4 "	2.37	2.64	3.03	3.21	3.52	3.75	4.09
5 "	2.46	2.73	3.08	3.27	3.66	3.91	4.31
6 "	2.56	2.78	3.11	3.29	3.69	4.07	4.53
7 "	2.63	2.81	3.14	3.32	3.77	4.27	4.77
8 "	2.72	2.85	3.17	3.51	4.04	4.59	5.17
9 "	2.82	2.94	3.26	3.69	4.30	4.91	5.57
10 "	2.90	3.01	3.39	3.86	4.53	5.23	5.94
11 "	2.98	3.10	3.53	4.08	4.78	5.57	6.38
12 "	3.07	3.22	3.70	4.29	5.09	5.92	6.80
13 "	3.14	3.30	3.88	4.51	5.38	6.29	7.23
14 "	3.18	3.43	4.07	4.72	5.66	6.63	7.66
15 "	3.25	3.56	4.23	4.95	5.94	7.00	8.09
16 "	3.33	3.69	4.40	5.18	6.24	7.35	8.52
17 "	3.38	3.82	4.57	5.39	6.52	7.70	8.95
18 "	3.47	3.94	4.74	5.61	6.80	8.06	9.37
19 "	3.59	4.10	4.92	5.82	7.10	8.42	9.80
20 "	3.71	4.23	5.09	6.05	7.38	8.76	10.23
21 "	3.84	4.36	5.27	6.28	7.66	9.13	10.66
22 "	3.94	4.49	5.44	6.49	7.95	9.48	11.09
23 "	4.05	4.61	5.61	6.71	8.24	9.83	11.51
24 "	4.15	4.74	5.78	6.93	8.53	10.19	11.95
25 "	4.25	4.86	5.95	7.15	8.80	10.55	12.37
26 "	4.35	4.99	6.13	7.38	9.10	10.89	12.80
27 "	4.45	5.14	6.31	7.59	9.38	11.26	13.24
28 "	4.55	5.26	6.48	7.81	9.67	11.61	13.65
29 "	4.64	5.39	6.65	8.04	9.96	11.97	14.10
30 "	4.74	5.52	6.84	8.28	10.28	12.36	14.57
31 "	4.83	5.69	7.05	8.54	10.60	12.74	15.04
32 "	4.92	5.85	7.24	8.80	10.93	13.14	15.51
33 "	5.01	5.99	7.45	9.05	11.25	13.53	15.98
34 "	5.09	6.14	7.63	9.26	11.55	13.91	16.45
35 "	5.19	6.26	7.79	9.49	11.84	14.31	16.89
36 "	5.27	6.41	7.97	9.72	12.13	14.65	17.34
37 "	5.36	6.55	8.15	9.94	12.42	15.03	17.77
38 "	5.45	6.68	8.32	10.17	12.72	15.38	18.21
39 "	5.53	6.81	8.51	10.38	13.01	15.74	18.64
40 "	5.62	6.93	8.67	10.61	13.29	16.11	19.07

Figure 5–3.

USING THE 3 CS

You have a package weighing 17 pounds 2 ounces. It is going to a business address in ZIP Code 34211. How much will it cost to send it by UPS?

COMPREHEND

In the space below, write what you are to do.

Find the rate for sending the package by UPS.

Write the steps you should follow to solve the problem.

1. Find the zone for the ZIP Code.
2. Find the rate in the Commercial Deliveries Rate Chart.

Write the information you need to solve the problem.

Business address; ZIP Code starts with 342; weight 17 pounds 2 ounces.

Write the system of measurement.

Weight and money

COMPUTE

Do the computation required to complete each task.

Zone is 6 for 342...
Rate for 17 pounds 2 ounces = 18 pound rate = $6.50

COMMUNICATE

Write the rate and shipping method as you would write them on the package.

$6.50 UPS

---

10 You have a package weighing 22 pounds going to a home address in ZIP Code 57700. How much will it cost to send it by UPS?

COMPREHEND

In the space below, write what you are to do.

_____

Write the steps you should follow to solve the problem.

_____

_____

Write the information you need to solve the problem.

_____

_____

Write the system of measurement.

_____

COMPUTE
Do the computation required to complete each task.

_____

_____

COMMUNICATE
Write the rate and shipping method as you would write them on the package.

_____

11 You have a package weighing 15 pounds 11 ounces going to a business address in ZIP Code 10011. How much will it cost to send it by UPS?

COMPREHEND
In the space below, write what you are to do.

_____

Write the steps you should follow to solve the problem.

_____

_____

Write the information you need to solve the problem.

_____

_____

Write the system of measurement.

_____

COMPUTE
Do the computation required to complete each task.

_____

_____

COMMUNICATE
Write the rate and shipping method as you would write them on the package.

_____

12 You have a package weighing 41 pounds 1 ounce going to a business address in ZIP Code 97521. How much will it cost to send it by UPS?

COMPREHEND
In the space below, write what you are to do.

_____

Write the steps you should follow to solve the problem.

_____

_____

Write the information you need to solve the problem.

_____

_____

Write the system of measurement.

_____

COMPUTE
Do the computation required to complete each task.

_____

_____

COMMUNICATE
Write the rate and shipping method as you would write them on the package.

_____

TASK 4: COMPARING FOURTH-CLASS AND UPS RATES

Your company usually prefers to send packages by UPS. However, it also wants to send packages in the least expensive way. So, you always compare UPS and fourth-class rates. If the difference between the two is less than $1, you choose UPS.

Here are the steps you follow:

1. Find the fourth-class rate.
2. Find the UPS rate.
3. Compare the rates and choose fourth class or UPS.
   ❑ If the UPS rate is less, choose UPS.
   ❑ If the fourth-class rate is less, subtract it from the UPS rate. If the difference is $1 or less, choose UPS. If the difference is greater than $1, choose fourth class.

USING THE 3 CS

You have a 17-pound package going to a business address in ZIP Code 88311. (This is Zone 8 for the U.S.P.S.) What shipping method should you use, and what is the cost?

COMPREHEND

In the space below, write what you are to do.

Find the best shipping method and the shipping cost.

Write the steps you should follow to solve the problem.

> 1. Find the fourth-class rate.
> 2. Find the UPS rate.
> 3. Compare the rates and choose fourth class or UPS.

Write the information you need to solve the problem.

> *Business address; ZIP Code 883...; Zone 8 U.S.P.S.; weight 17 pounds.*

Write the system of measurement.

> *Weight and money*

COMPUTE
Do the computation required to complete each task.

> 1. *17 pounds in Zone 8 = $19.91 fourth class*
> 2. *Zone = 8 UPS*
>   *17 pounds in Zone 8 = $8.65 (commercial rate)*
> 3. *The UPS rate is less. Choose UPS.*

COMMUNICATE
Write the rate and type of mail as you would write them on the package.

> *$8.65 UPS*

PRACTICE

13 You have a 69-pound package going to a business address in ZIP Code 33802. (This is Zone 5 for the U.S.P.S.) What shipping method should you use, and what is the cost?

COMPREHEND
In the space below, write what you are to do.

_____

Write the steps you should follow to solve the problem.

_____

_____

_____

Write the information you need to solve the problem.

_____

_____

Write the system of measurement.

_____

COMPUTE
Do the computation required to complete each task.

_____

_____

_____

_____

COMMUNICATE
Write the rate and type of mail as you would write them on the package.

_____

14 You have a 10-pound 14-ounce package going to a business address in ZIP Code 88107. (This is Zone 7 for the U.S.P.S.) What shipping method should you use, and what is the cost?

COMPREHEND
In the space below, write what you are to do.

_____

Write the steps you should follow to solve the problem.

_____

_____

_____

Write the information you need to solve the problem.

_____

_____

Write the system of measurement.

_____

COMPUTE
Do the computation required to complete each task.

_____

_____

_____

COMMUNICATE

Write the rate and type of mail as you would write them on the package.

_____

15 You have a 40-pound package going to a business address in ZIP Code 77822. (This is Zone 6 for the U.S.P.S.) What shipping method should you use, and what is the cost?

COMPREHEND

In the space below, write what you are to do.

_____

Write the steps you should follow to solve the problem.

_____

_____

_____

Write the information you need to solve the problem.

_____

_____

Write the system of measurement.

_____

COMPUTE

Do the computation required to complete each task.

_____

_____

_____

_____

COMMUNICATE

Write the rate and type of mail as you would write them on the package.

_____

16 You have a 45-pound package going to a business address in ZIP Code 47555. (This is Zone 5 for the U.S.P.S.) What shipping method should you use, and what is the cost?

COMPREHEND

In the space below, write what you are to do.

_____

Write the steps you should follow to solve the problem.

_____

_____

_____

Write the information you need to solve the problem.

_____

_____

Write the system of measurement.

_____

COMPUTE
Do the computation required to complete each task.

_____

_____

_____

_____

COMMUNICATE
Write the rate and type of mail as you would write them on the package.

_____

TASK 5: USING OVERNIGHT MAIL

Overnight mail gets a letter or package to its destination quickly. You use it when someone asks for overnight service for a package.

Sending packages overnight is expensive. The quicker the package has to be delivered, the more it costs. Figure 5-4 shows some rates for letters and small packages from an overnight mail service.

	Priority Service (by 10:30 A.M. next day)	Next Day Service (by 3:00 P.M. next day)	Second Day Service (by 3:00 P.M. second day)
Letter (up to 8 ounces)	$12.50	$9.50	$7.95
Package - up to 2 pounds	$15.50	$12.50	$9.95
- 2 to 5 pounds	$20.50	$17.00	$14.95
- 5 to 7 pounds	$22.50	$19.50	$15.50

Figure 5-4 Overnight Mail Rates

When you send a package by overnight mail, you fill out the **air bill** for the package. The air bill is the shipper's form that tells where the package is going. You also write on the air bill the kind of delivery service you want. When you are given a package for overnight mail, you always ask when the package has to arrive. Must it be there in the morning? Can it arrive after 3:00 p.m.? Can it arrive the following day? Then, when you fill out the air bill, choose the least expensive service that will get the package there in time. Once you know the type of service and the weight of the package, you can use the rate table to find the delivery cost.

USING THE 3 CS

You are given a package that must be delivered by noon tomorrow. The package weighs 5 ounces. What delivery service will you use, and what will it cost?

COMPREHEND
In the space below, write what you are to do.

> *Pick the type of delivery service and find the cost.*

Write the steps you should follow to solve the problem.

> *1. Choose the latest possible delivery.*
> *2. Find the rate.*

Write the information you need to solve the problem.

> *Delivery by noon; weight 5 ounces.*

Write the system of measurement.

> *Weight and money*

COMPUTE
Do the computation required to complete each task.

> *To get there by noon, you need to use Priority Service.*
> *Priority Service rate for 5 ounces is $12.50.*

COMMUNICATE
Write the rate. Write the type of service as you would write it on the air bill.

> *$12.50 Priority Service*

---

17 You are given a package that must be delivered by 5:00 p.m. tomorrow. The package weighs 3 pounds. What delivery service will you use, and what will it cost?

COMPREHEND
In the space below, write what you are to do.

_____

Write the steps you should follow to solve the problem.

_____

_____

Write the information you need to solve the problem.

_____

Write the system of measurement.

_____

COMPUTE
Do the computation required to complete each task.

_____

_____

COMMUNICATE
Write the rate. Write the type of service as you would write it on the air bill.

_____

18 You are given a package that must arrive by 4 p.m. the day after tomorrow. The package weighs 8 ounces. What delivery service will you use, and what will it cost?

COMPREHEND
In the space below, write what you are to do.

_____

Write the steps you should follow to solve the problem.

_____

_____

Write the information you need to solve the problem.

_____

Write the system of measurement.

_____

COMPUTE
Do the computation required to complete each task.

_____

_____

COMMUNICATE
Write the rate. Write the type of service as you would write it on the air bill.

_____

19 You are given a package that must arrive by 1 p.m. tomorrow. The package weighs 6 pounds. What delivery service will you use, and what will it cost?

COMPREHEND
In the space below, write what you are to do.

_____

Write the steps you should follow to solve the problem.

_____

_____

Write the information you need to solve the problem.

_____

Write the system of measurement.

_____

COMPUTE
Do the computation required to complete each task.

_____

_____

COMMUNICATE
Write the rate. Write the type of service as you would write it on the air bill.

_____

❏ JOB SITUATION 2 THE COPY CENTER

People who work in the Copy Center at Gibson are responsible for copying all kinds of documents. In the Copy Center, you do work for all the other departments in the company. Each department sends documents to copy. The department gives you a requisition form like the one shown in Figure 7-1 of the *Office Technology Knowledge Base*.

In the Copy Center, you do the work asked for on the form. The form tells you how many copies to make. It also tells you the number of pages in the original document. Sometimes you need to make the copy larger or smaller than the original. You need to bind the document in some way. Sometimes you just staple the copies. Other times you need to do special binding such as a comb binding.

At Gibson, the Copy Center keeps track of the amount of work it does for other departments. Each time you do work for a department at Gibson, that department is charged for the work. Each time you complete a job, you fill out a log telling whom the work was for. You find out the cost, using the Copy Center Rate Chart, and record that cost in the log.

TASK 1: MAKING STANDARD REDUCTIONS AND ENLARGEMENTS

When you need to make a reduction or enlargement, you need to know what percentage to use. The chart on page 86 in the *Knowledge Base* shows what percent reductions to use to reduce some standard sizes to 8½- by 11-inch copies. Find the original size and the copy size in the chart to find the percent reduction to use.

When you need to make enlargements, you can also use the chart in the *Knowledge Base* to find what percent enlargement to use.

USING THE 3 Cs

The requisition shown in Figure 7-6 in the *Knowledge Base* says that you should reduce a drawing and include it in the report. The drawing is 8½ by 14 inches. What percent should you use on the copy machine?

COMPREHEND
In the space below, write what you are to do.

Find the percent to use on the copy machine.

Write the step you should follow to solve the problem.

Look up the original and copy size in the chart.

Write the information you need to solve the problem.

The original is 8½ by 14 inches. The copy is to be 8½ by 11.

Write the system of measurement.

Length and width

COMPUTE
Do the computation required to complete each task.

The chart shows that the reduction is 77% of the original.

COMMUNICATE
Write the percent as you would write it on a self-sticking note to attach to the drawing.

77% reduction

PRACTICE

1 You get a requisition for a document that needs to be reduced to 8½- by 11-inch paper. The document is 11 by 15 inches. What percent should you use on the copy machine?

COMPREHEND
In the space below, write what you are to do.

_____

Write the step you should follow to solve the problem.

_____

Write the information you need to solve the problem.

_____

_____

Write the system of measurement.

_____

COMPUTE
Do the computation required to complete each task.

_____

COMMUNICATE
Write the percent reduction as you would write it on a self-sticking note to attach to the drawing.

_____

2 You get a requisition for a document that needs to be enlarged to 11- by 17-inch paper. The document is 8½ by 11 inches. What percent enlargement do you need?

COMPREHEND
In the space below, write what you are to do.

_____

Write the step you should follow to solve the problem.

_____

Write the information you need to solve the problem.

_____

_____

Write the system of measurement.

_____

COMPUTE
Do the computation required to complete each task.

_____

COMMUNICATE
Write the percent enlargement as you would write it on a self-sticking note to attach to the drawing.

_____

TASK 2: MAKING NONSTANDARD REDUCTIONS AND ENLARGEMENTS

Some copy machines can make more than three types of reductions and enlargements. This kind of machine might be able to reduce to any percentage from 50 to 99 percent of the original size. It might be able to enlarge anywhere from 101 to 200 percent of the original size. You would use this kind of machine to make reductions and enlargements when you can't find the original and copy size on the charts.

When you need a very high quality reduction or enlargement, you would send the original out to a service that makes photostats. Photostats are photos of the document. They are much clearer and sharper than any photocopy you can make. You need to know what percent reduction or enlargement to ask for when you make a phototstat.

To choose the percent on the machine or to ask for the percent from a photostat service, use the percent equation. Solve the percent equation to find the percent:

$$\text{percent} \times \text{base} = \text{amount}$$
$$(\text{percent} \times \text{original size} = \text{new size})$$

You may also need to know the length and width of the copy. This can be helpful if you need to know whether the new copy will fit in a blank space on a page. For example, if you are reducing a page that is 8½ by 11 to a copy 4¼ wide, you know the new width. To find the new length, solve a proportion:

$$\frac{\text{original width}}{\text{original length}} = \frac{\text{new width}}{\text{new length}} \quad \leftarrow \text{ This is the number you need to find.}$$

USING THE 3 CS

You have a requisition asking that you reduce a form that is 8½ by 11 inches so that it is 4¼ inches wide. What percent should you use for reducing on the copy machine? If the copy is 4¼ inches wide, how long will it be?

COMPREHEND
In the space below, write what you are to do.

Find the percent to use and find the new length.

Write the steps you should follow to solve the problem.

1. Solve an equation to find the percent.
2. Solve a proportion to find the length.

Write the information you need to solve the problem.

original width = 8½ inches
original length = 11 inches
new width = 4¼ inches

Write the system of measurement.

Length and width

COMPUTE

Do the computation required to complete each task.
Use the formulas given on page 81.

$$percent \times base = amount$$
$$(percent \times original\ size = new\ size)$$

n	$\times$	$8\frac{1}{2}$	$=$	$4\frac{1}{4}$
n	$\times$	8.5	$=$	4.25
	$\div$	8.5	$\div$	8.5
		n	$=$	$.50$
			50%	

$$\frac{original\ width}{original\ length} = \frac{new\ width}{new\ length}$$

$$\frac{8.5}{11} = \frac{4.25}{n} \qquad \begin{aligned} 8.5 \times n &= 46.75 \\ \div 8.25 & \div 8.25 \\ n &= 5.67\ inches \end{aligned}$$

COMMUNICATE

Write the percent that you will use in copying and the new length.

50%   5.67 inches

PRACTICE

3 You have a requisition asking that you reduce a photo that is 6 by 8 inches so that it is 3 inches wide. What percent should you ask for from the photostat service? How long will the copy be?

COMPREHEND

In the space below, write what you are to do.

_____

Write the steps you should follow to solve the problem.

_____

_____

Write the information you need to solve the problem.

_____

_____

_____

Write the system of measurement.

_____

COMPUTE
Do the computation required to complete each task.
Use the formulas given on page 81.

COMMUNICATE
Write the percent that you will use in copying and the new length.

_____

4 You have a requisition asking that you reduce a page that is 6 by 9 inches so that it is 4½ inches wide. What percent should you use? How long will the copy be?

COMPREHEND
In the space below, write what you are to do.

_____

Write the steps you should follow to solve the problem.

_____

_____

Write the information you need to solve the problem.

_____

_____

_____

Write the system of measurement.

_____

COMPUTE
Do the computation required to complete each task.
Use the formulas given on page 81.

COMMUNICATE

Write the percent that you will use in copying and the new length.

_____

5 You have a requisition asking that you reduce a photo that is 8 by 10 inches so that it is 2¾ inches wide. What percent should you ask for from the photostat service? How long will the copy be?

COMPREHEND

In the space below, write what you are to do.

_____

Write the steps you should follow to solve the problem.

_____

_____

Write the information you need to solve the problem.

_____

_____

Write the system of measurement.

_____

COMPUTE

Do the computation required to complete each task.
Use the formulas given on page 81.

COMMUNICATE

Write the percent that you will use and the new length.

_____

6 You have a requisition asking that you enlarge a photo that is 5 by 7 inches so that it is 8 inches wide. What percent should you ask for from the photostat service? How long will the copy be?

Part 5: Mathematics and Material Resources

COMPREHEND

In the space below, write what you are to do.

_____

Write the steps you should follow to solve the problem.

_____

Write the information you need to solve the problem.

_____

_____

Write the system of measurement.

_____

COMPUTE

Do the computation required to complete each task.
Use the formulas given on page 81.

COMMUNICATE

Write the percent that you will use and the new length.

_____

7 You have a requisition asking that you enlarge a map that is 10½ by 12 inches so that it is 15¾ inches wide. What percent should you ask for from the photostat service? How long will the copy be?

COMPREHEND

In the space below, write what you are to do.

_____

Write the steps you should follow to solve the problem.

_____

_____

Write the information you need to solve the problem.

_____

_____

_____

Write the system of measurement.

_____

COMPUTE
Do the computation required to complete each task.
Use the formulas given on page 81.

COMMUNICATE
Write the percent that you will use and the new length.

_____

TASK 3: FILLING OUT THE COPY LOG

In order to keep track of work for each department, you fill out a log like the one shown in Figure 5-5. It shows the number of copies you made. It shows whether you used a special binding or other special service. It also shows the total cost.

To calculate the cost for copies, you need to use the Gibson Copy Rate Chart. (See Figure 5-6.) This shows the cost per copy. Notice that the cost is more for double-sided copies (copies that are printed on both sides of the page). The cost is less when offset duplication is used. The cost for an offset copy depends on the total number of copies made.

Date: September 23

Department	Total Copies	Special Services	Total Cost of Copies
Acct.	60	2-sided	$6.00

Figure 5-5 Copy Log

```
┌─────────────────────────────────────────────────────┐
│              RATES FOR COPY SERVICES                  │
│                                                       │
│                      COPIES                           │
│   Regular Copies                    7¢ per copy       │
│   Double-Sided Copies               10¢ per copy      │
│   Legal-Size Copies                 8¢ per copy       │
│                                                       │
│                 OFFSET DUPLICATION                    │
│   1 - 499 total copies              6¢ per copy       │
│   500 - 4,999 total copies          4.5¢ per copy     │
│   5,000 and up total copies         2.5¢ per copy     │
│                                                       │
│                     BINDING                           │
│   Stapling                          No charge         │
│   Comb Binding (spiral)             25¢ each          │
│                  (flat)             10¢ each          │
└─────────────────────────────────────────────────────┘
```

Figure 5-6 Gibson Copy Rate Chart

To find the cost for copies:

1. Multiply the number of originals times the number of copies to find the total number of copies.
2. Find the rate per copy on the chart. Multiply the rate times the total number of copies.
3. Add the cost for any extra services.

USING THE 3 Cs

You make 50 copies each of 7 pages for the Data Processing Department. You use spiral comb binding on each set. Find the total cost of the copies.

COMPREHEND

In the space below, write what you are to do.

Find the cost of the copies.

Write the steps you should follow to solve the problem.

1. Find the total number of copies.
2. Multiply the rate times the total number of copies.
3. Add the cost for any extra services.

Write the information you need to solve the problem.

Data Processing Department, 50 copies, 7 pages each, spiral binding

Write the system of measurement.

Money

COMPUTE
Do the computation required to complete each task.
Use the formulas given on page 87.

50 copies × 7 pages = 50 × 7 = 350
350 copies × 7¢ a copy = 350 × .07 = $24.50
50 spiral binders × 25¢ a binder = 50 × .25 = $12.50
copy cost + special service cost = $24.50 + $12.50 = $37

COMMUNICATE
Write the department, number of copies, special services, and the cost on the copy log below.

Date: September 24

Department	Total Copies	Special Services	Total Cost of Copies
DP	350	spiral bind 50 copies	$37.00

PRACTICE

8 You make 10 copies each of 5 pages for the Accounting Department. Find the total cost of the copies.

COMPREHEND
In the space below, write what you are to do.

_____

Write the steps you should follow to solve the problem.

_____

_____

_____

Write the information you need to solve the problem.

_____

_____

Write the system of measurement.

_____

COMPUTE
Do the computation required to complete each task.
Use the formulas given on page 87.

_____
_____
_____
_____

COMMUNICATE
Write the department, number of copies, special services, and the cost
on the copy log on page 88.

9 You make 20 copies each of 14 pages for the Accounting Department.
The copies are on legal-size paper. Find the total cost of the copies.

COMPREHEND
In the space below, write what you are to do.

_____

Write the steps you should follow to solve the problem.

_____
_____
_____

Write the information you need to solve the problem.

_____
_____

Write the system of measurement.

_____

COMPUTE
Do the computation required to complete each task.
Use the formulas given on page 87.

_____
_____
_____

COMMUNICATE
Write the department, number of copies, special services, and the cost
on the copy log on page 88.

10 You make 30 copies each of 27 pages for the Sales Department. You bind them with flat comb binding. Find the total cost of the copies.

COMPREHEND
In the space below, write what you are to do.

_____

Write the steps you should follow to solve the problem.

_____

_____

_____

Write the information you need to solve the problem.

_____

_____

Write the system of measurement.

_____

COMPUTE
Do the computation required to complete each task.
Use the formulas given on page 87.

_____

_____

_____

_____

COMMUNICATE
Write the department, number of copies, special services, and the cost on the copy log on page 88.

11 You make 200 copies each of 30 pages for the Sales Department on the offset duplicating machine. Find the total cost of the copies.

COMPREHEND
In the space below, write what you are to do.

_____

Write the steps you should follow to solve the problem.

_____

_____

_____

Write the information you need to solve the problem.

_____

_____

Write the system of measurement.

_____

COMPUTE
Do the computation required to complete each task.
Use the formulas given on page 87.

_____

_____

_____

COMMUNICATE
Write the department, number of copies, special services, and the cost on the copy log on page 88.

12 You make 100 copies each of 25 pages for the Data Processing Department on the offset duplicating machine. You use spiral comb binding. Find the total cost of the copies.

COMPREHEND
In the space below, write what you are to do.

_____

Write the steps you should follow to solve the problem.

_____

_____

_____

Write the information you need to solve the problem.

_____

_____

Write the system of measurement.

_____

COMPUTE
Do the computation required to complete each task.
Use the formulas given on page 87.

_____

_____

_____

_____

COMMUNICATE
Write the department, number of copies, special services, and the cost on the copy log on page 88.

❏ JOB SITUATION 3 TRAVEL ARRANGEMENTS

When you work as an office assistant, one of your jobs is making travel arrangements for your supervisor. If your supervisor is flying, you are responsible for getting airline reservations and tickets. You pick the flights. When you pick the flights, you need to know when your supervisor needs to fly. For example, there may be a meeting in the morning, and your supervisor can't leave for the airport until noon. You know that it can take an hour to get to the airport. So you won't pick a flight that leaves before 1:30 or 2 p.m.

Once you have picked flights, you check with your supervisor. Then you may call the airline yourself to make reservations. Or you may use a travel agent to make the reservations.

If your supervisor is flying or taking the train to another city, always check to see if a rental car is needed there. Sometimes it's convenient to take cabs to get around a city. But if there are a number of different stops or if the places to be visited are in the suburbs, it is probably necessary to have a car. Ask your supervisor what size of car will be needed.

When your supervisor needs a rental car, you can reserve it through a travel agent or by calling the car rental agency directly. Your supervisor will want to pick up the car at the airport. You tell the car rental agency the date and time that the flight arrives so that they will have a car ready.

When you reserve a car, you will find out the rate the car rental agency charges. Typical rates are shown in Figure 4-5 in the *Knowledge Base*. These rates are daily rates based on the size of the car. The company will also charge for the number of miles driven. There is a separate daily charge for insurance. There may be an extra charge for dropping the car off at a different location from where it was rented. Once you find out the rates, you estimate how much the rental car will cost. Your supervisor will probably have some idea of the number of miles he or she will drive, so you can use that number in your estimate.

Sometimes car rental agencies offer special weekly or daily rates. These usually include either unlimited free miles or a very large number of miles. When you estimate the cost for the rental car, you may find out that the special rate will be cheaper. For example, say you estimate that the cost of a rental car for four days will be $200. In that case, it would be better to take a five-day special for $150.

When the trip is over, your company will receive a bill from the car rental agency. It will show the actual miles driven. You check the bill, using the rates you were given and the actual mileage.

TASK 1: SELECTING FLIGHTS

As an office assistant, you may use the *Official Airline Guide (OAG)* to help you see what flights are available. An example from the *OAG* is shown in Figure 4-3 in the *Knowledge Base*.

The *OAG* is arranged by destination cities. Figure 4-3 shows part of the guide for the destination Dallas. This is where you would look if your supervisor was flying to Dallas. Under the destination are tables for flights from all the cities with flights to Dallas. Figure 4-3 shows the table of flights from Chicago to Dallas.

Under the Chicago heading are all the flights from Chicago to Dallas. The first group of flights are direct flights. You stay on one plane the whole time. Then below that are connecting flights. These are flights where you fly to one city, get off the plane, change to another plane, and continue to the destination. People prefer to fly directly when they can.

For each of the flights, there are listings of flight times. Each shows the time that a plane will leave Chicago and arrive in Dallas. There are codes for the airports. Both Dallas and Chicago have more than one airport. Be sure to pick right airport!

There is a code for the airline and a flight number for the flight. Figure 5-7 shows what each item in a line in the *OAG* means.

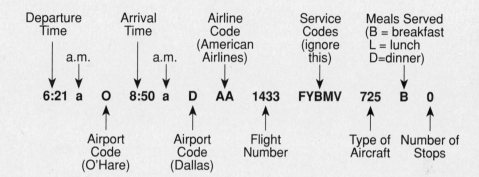

Figure 5-7

To pick a flight:

❑ Look for a flight at or after the time your supervisor wants to fly. Write the time, airline, and flight number. Use the index shown in Figure 5-8 to tell what the airline code means.

❑ Check to be sure the flight operates on the day you want.

If there is a code starting with an *X*, for example, *X7*, at the beginning of the line, the flight does not operate on the seventh day—Sunday. (Mon. = 1, Tues. = 2, Wed. = 3, Thurs. = 4, Fri. = 5, Sat. = 6, Sun. = 7)

If there is a number code, for instance, 6, the flight operates only on the sixth day—Saturday.

If there is no code, the flight operates every day.

❑ If you have to pick a connecting flight, write down the flight numbers, airlines, times, and cities for the whole flight.

Airline Code	Airline
AA	American Airlines
CO	Continental Airlines
HP	Air West Airlines
NW	Northwest Airlines
UA	United Airlines

City Code	City
IAH	Houston
MSP	Minneapolis

Figure 5-8

USING THE 3 CS

Your supervisor wants to fly from Chicago's O'Hare Airport to New Orleans on Monday around 7:30 a.m. Then she wants to fly from New Orleans to Phoenix on Tuesday after 6 p.m. Finally, she wants to return on Friday from Phoenix to O'Hare Airport, leaving around 10 a.m. Pick the flights you would suggest to her. Write down the airlines, times, and flight numbers. (Use the portion of the *OAG* shown here.)

To NEW ORLEANS, LOUISIANA CDT MSY
M-MSW (INTERNATIONAL)
N-NEW (LAKEFRONT)

▲CHICAGO, ILLINOIS
M-MDW O-ORD C-CGX P-PWK

X67	6:15a	M	9:25a	M	WN	443	YQ	733	2
X7	6:20a	O	9:49a	M	NW	471	FYBQM	72S	S	1
	7:44a	O	9:51a	M	UA	463	FYBMQ	727	B	0
	8:40a	M	10:48a	M	ML	346	YBHQM	D9S	S	0
	1:50P	O	3:59P	M	UA	721	FYBMQ	727	S	0
	2:15P	M	4:23P	M	ML	044	YBHQM	D9S	S	0
	4:45P	M	8:35P	M	WN	758	YQ	73S	2
	4:46P	M	8:50P	M	TW	385	FYMBQ	727	•	1

TW 385 • MEALS SD/S

	6:45P	O	9:02P	M	UA	469	FYBMQ	727	D	0
X6	7:45P	M	9:53P	M	ML	356	YBHQM	73S	S	0

ML 356 EFFECTIVE 15AUG
CONNECTIONS

6:00a	O	7:09a	STL	TW	531	FYMBQ	09S	S/	0
				TW	539	FYMBQ	09S	B	0

Table 5-1

To PHOENIX, ARIZONA MST PHX
ALSO SEE SCOTTSDALE, ARIZONA

▲NEW ORLEANS, LOUISIANA CDT MSY
M-MSY N-NEW

6:15a	M	8:51a	AA	923	FYBMV	M80	•	1

AA 923 DISCONTINUED AFTER 30AUG
AA 923 • MEALS B/SB

6:15a	M	8:53a	AA	923	FYBMV	M80	•	1

AA 923 EFFECTIVE 31AUG
AA 923 • MEALS B/SB

11:40a	M	1:35P	WN	743	YQ	733	1

CONNECTIONS

4:55P	M	7:03P	MDW	ML	341	YBHQM	09S	S	0
D-14AUG 8:30P	MDW	10:10P		ML	762	YBHQM	73S	S	0
6:40P	M	7:52P	IAH	CO	892	FYBMQ	D9S		0
8:30P	IAH	9:27P		CO	1075	FYBMQ	M80	S/	0
7:41P	M	9:04P	DFW	AA	145	FYBMV	767		0

Table 5-2

To CHICAGO, ILLINOIS CDT CHI
▲PHOENIX, ARIZONA MST PHX

	12:35a	5:45a	O	AA	890	FnYnBMV	M80	0

AA 890 EFFECTIVE 31AUG

	12:35a	5:52a	O	AA	890	FnYnBMV	M80	0

AA 890 DISCONTINUED AFTER 30AUG

	12:40a	6:00a	M	ML	302	YBHQM	73S	0

ML 302 DISCONTINUED AFTER 14AUG

	4:40a	11:35a	O	CO	80	FYnBMQ	72S	SB	1
X7	5:35a	12:00n	M	WN	668	YQ	733	2
X67	6:45a	2:00P	M	WN	922	YQ	73S	2
X67	7:00a	12:20P	O	HP	2	FYKQV	757	B	0
	7:05a	12:19P	O	AA	550	FYBMV	767	B	0

AA 550 EFFECTIVE 31AUG

	7:05a	12:27P	O	AA	550	FYBMV	767	B	0

AA 550 DISCONTINUED AFTER 30AUG

	7:35a	1:05P	M	UA	772	FYBMQ	733	B	0
X67	8:00a	6:05P	M	WN	516	YQ	73S	5
	8:30a	2:30P	M	WN	702	YQ	733	1
	9:52a	3:09P	O	AA	242	FYBMV	M80	L	0

AA 242 EFFECTIVE 31AUG

	9:52a	3:11P	O	AA	242	FYBMV	M80	L	0

AA 242 DISCONTINUED AFTER 30AUG

	10:00a	3:14P	O	HP	4	YKMQV	733	S	0
	11:00a	4:06P	O	UA	266	FYBMQ	72S	L	0
	11:00a	4:20P	M	ML	310	YBHQM	73S	S	0

ML 310 DISCONTINUED AFTER 14AUG

	11:00a	4:20P	M	ML	350	YBHQM	73S	0

ML 350 EFFECTIVE 15AUG

	11:30a	4:435P	O	AA	434	FYBMV	M80	L	0

AA 434 DISCONTINUED AFTER 30AUG

	11:30a	4:47P	O	AA	434	FYBMV	M80	L	0

AA 434 EFFECTIVE 31AUG

	1:00P	8:10P	O	DL	906	FYBMQ	72S	SD	1
	1:20P	7:00P	O	UA	492	FYBMQ	733	O	0
	1:34P	7:00P	O	AA	918	FYBMV	767	L	0

Table 5-3

COMPREHEND

In the space below, write what you are to do.

Pick the flights for this trip.

Write the steps you should follow to solve the problem.

1. Look for a flight at or after the time your supervisor wants to fly.
2. Check to be sure the flight operates on the day you want.

Write the information you need to solve the problem.

1. Chicago's O'Hare Airport to New Orleans on Monday around 7:30 a.m.
2. New Orleans to Phoenix on Tuesday after 6 p.m.
3. Phoenix to O'Hare Airport on Friday around 10 a.m.

Write the system of measurement.

Time

COMPUTE

Do the computation required to complete each task.

1. Table 5-1: OAG shows flight at 7:44 a.m., arriving 9:51 a.m., all days, United Airlines, flight number 463.

2. Table 5-2: OAG shows connecting flight at 6:40 p.m. arriving 7:52 p.m., in Houston (IAH), Continental Airlines, flight number 892, all days; flight leaving Houston at 8:30 p.m., arriving 9:27 p.m. in Phoenix, Continental Airlines, flight number 1075.
3. Table 5-3: OAG shows flight at 10 a.m., arriving 3:14 p.m., all days, Air West Airlines, flight number 4.

COMMUNICATE
Write the flight numbers, times, and airlines as you would put them in a note to your supervisor.

> *Monday: Chicago to New Orleans, depart 7:44 a.m., arrive 9:51 a.m., United Airlines 463.*

> *Tuesday: New Orleans to Phoenix via Houston:*

> ❏ *New Orleans to Houston, depart 6:40 p.m., arrive 7:52 p.m., Continental Airlines 892.*

> ❏ *Houston to Phoenix, depart 8:30 p.m., arrive 9:27 p.m., Continental Airlines 1075.*

> *Friday: Phoenix to Chicago, depart 10 a.m., arrive 3:14 p.m., Air West Airlines 4.*

PRACTICE
1 Your supervisor wants to fly from Chicago to Dallas on Monday. Use the portion of the *OAG* shown in Figure 4-3 in the *Knowledge Base* to pick a flight around 3 p.m. Write the airline, times, and flight number.

COMPREHEND
In the space below, write what you are to do.

_____

Write the steps you should follow to solve the problem.

_____

_____

Write the information you need to solve the problem.

_____

Write the system of measurement.

_____

COMPUTE
Do the computation required to complete each task.

_____

_____

COMMUNICATE

Write the flight numbers, times, and airlines as you would put them in a note to your supervisor.

_____

_____

2 Your supervisor wants to fly from Seattle to Kansas City on Tuesday. Pick a flight after noon. Write the airline, times, and flight number.

To KANSAS CITY, MISSOURI			CDT	MKC
▲SEATTLE, WASHINGTON			PDT	SEA
S-SEA B-BFI L-LKE E-PAE				

7:30a	S	12:00P	O	EA	200	FYHQK	72S	B	0
11:55a	S	5:07P	C	BN	582	YQLMK	72S	S	0
12:45P	S	5:55P	C	NW	66	FYHQK	72S	L	0
4:30P	S	11:20P	C	DL	1892	FYBMQ	72S	•	1
				DL1892 • MEALS DS/D				

Table 5-4

COMPREHEND

In the space below, write what you are to do.

_____

Write the steps you should follow to solve the problem.

_____

_____

_____

Write the information you need to solve the problem.

_____

Write the system of measurement.

_____

COMPUTE

Do the computation required to complete each task.

_____

_____

COMMUNICATE

Write the flight numbers, times, and airlines as you would put them in a note to your supervisor.

_____

_____

3 Your supervisor wants to fly from Chicago to Denver, Colorado, around 11 a.m. Wednesday. Pick the flight. Write the airline, times, and flight number.

To DENVER, COLORADO						MDT	DEN
D-DEN (STAPLEDON)							
▲CHICAGO, ILLINOIS						CDT	CHI
M-MDW O-ORD C-CGX P-PWK							
6:20a O	7:46a D	AA	299	FYBMQ	D10	B	0
6:40a M	8:07a D	UA	455	FYBMQ	73S	B	0
6:40a M	8:11a D	CO	881	FYBMQ	727	B	0
6:40a O	8:13a D	CO	63	FYBMQ	AB3	B	0
8:00a O	9:29a D	UA	221	FYBMQ	72S	B	0
8:40a M	10:10a D	ML	782	YBHQM	DC9	S	0
9:00a O	10:35a D	UA	223	FYBMQ	733	B	0
9:50a O	11:25a D	CO	83	FYBMQ	72S	B	0
10:00a O	11:28a D	UA	225	FYBMQ	D10	S	0
11:00a O	12:18P D	UA	227	FYBMQ	D8S	L	0
12:00n O	1:20P D	UA	229	FYBMQ	727	L	0
12:20P O	1:42P D	CO	67	FYBMQ	72S	L	0
12:50P O	4:18P D	UA	959	FYBMQ	733	•	1
UA 959 • MEALS LS/L							

Table 5-5

COMPREHEND
In the space below, write what you are to do.

_____

Write the steps you should follow to solve the problem.

_____

_____

_____

Write the information you need to solve the problem.

_____

Write the system of measurement.

_____

COMPUTE
Do the computation required to complete each task.

_____

_____

COMMUNICATE
Write the flight numbers, times, and airlines as you would put them in a note to your supervisor.

_____

_____

4 Your supervisor wants to fly to Madison, Wisconsin, on Thursday around 7 a.m. and return the same day around 7 p.m. Pick the flights. Write the airlines, times, and flight numbers.

To MADISON, WISCONSIN						CDT MSN
▲CHICAGO, ILLINOIS						CDT CHI
	M-MDW	O-ORD	C-CGX	P-PWK		
X67	7:00a	O	7:55a	AA★ 4221	YBMVQ	SH6	0
X7	7:44a	O	8:29a	UA★ 2841	YBMQH	F27	0
6	8:45a	O	9:35a	AA★ 4187	YBMVQ	SH6	0
X6	8:45a	O	9:40a	AA★ 4206	YBMVQ	SH6	0
X7	8:53a	M	9:43a	ML★ 1600	YBHQM	DO8	0
X6	10:20a	O	11:05a	UA★ 2845	YBMQH	F27	0
	11:30a	O	12:25P	AA★ 4255	YBMVQ	SH6	0
	11:38a	M	12:28P	ML★ 1674	YBHQM	DO8	0
	11:55a	O	12:40P	UA★ 2847	YBMQH	F27	0
	1:14P	O	1:59P	UA★ 2849	YBMQH	F27	0
	1:14P	O	2:09P	AA★ 4160	YBMVQ	SH6	0
	2:00P	M	2:50P	ML★ 1676	YBHQM	DO8	0
	2:44P	O	3:34P	AA★ 4257	YBMVQ	ATR	0
	3:47P	O	4:32P	UA★ 2851	YBMQH	F27	0
X6	4:00P	O	4:50P	AA★ 4242	YRMVQ	SH6	0
6	4:05P	O	4:55P	AA★ 4210	YBMVQ	SH6	0
	4:15P	M	5:05P	ML★ 1863	YBHQM	DO8	0
	4:44P	O	5:29P	UA★ 2857	YBMQH	F27	0
X6	5:25P	O	6:20P	AA★ 4251	YBMVQ	SH6	0
	6:45P	O	7:40P	AA★ 4166	YBMVQ	SH6	0
	7:15P	O	8:00P	UA★ 2859	YBMQH	F27	0
X6	7:36P	M	8:26P	ML★ 1688	YBHQM	DO8	0
X6	9:45P	O	10:21P	UA★ 2933	YBMQH	146	0
X6	9:45P	O	10:40P	AA★ 4180	YnBMVQ	SH6	0

Table 5-6

To CHICAGO, ILLINOIS						CDT CHI
C-CGX (MEIGS FIELD)						
M-MDW (MIDWAY)						
O-ORD (O'HARE)						
P-PWK (PAL-WAUKEE)						
▲MADISON, WISCONSIN						CDT MSN
X67	5:20a		6:10a	O	AA★ 4178	YnBMVQ	SH6	0
X7	6:30a		7:20a	M	ML★ 1769	YBHQM	DO8	0
X7	6:55a		7:35a	O	UA★ 2840	YBMQH	F27	0
	8:20a		9:15a	O	AA★ 4250	YBMVQ	SH6	0
	8:25a		9:03a	O	UA★ 2922	YBMQH	146	0
	9:35a		10:15a	O	UA★ 2842	YBMQH	F27	0
X6	10:00a		10:50a	O	AA★ 4234	YBMVQ	SH6	0
	10:10a		11:00a	M	ML★ 1779	YBHQM	DO8	0
	10:35a		11:15a	O	UA★ 2844	YBMQH	F27	0
X6	11:34a		12:14P	O	UA★ 2846	YBMQH	F27	0
X7	11:43a		12:38P	O	AA★ 4157	YBMVQ	CIIC	0
7	11:45a		12:40P	O	AA★ 4157	YBMVQ	SH6	0
	12:38P		1:23P	M	ML★ 1789	YBHQM	DO8	0
	1:04P		1:44P	O	UA★ 2848	YBMQH	F27	0
	1:20P		2:10P	O	AA★ 4179	YBMVQ	SH6	0
	2:30P		3:20P	O	AA★ 4173	YBMVQ	SH6	0
	3:00P		3:50P	M	ML★ 1659	YBHQM	DO8	0
	3:55P		4:44P	O	AA★ 4258	YBMVQ	ATR	0
	4:04P		4:44P	O	UA★ 2850	YBMQH	F27	0
	5:00P		5:40P	O	UA★ 2852	YBMQH	F27	0
X6	5:10P		6:00P	O	AA★ 4230	YBMVQ	SH6	0
X6	5:15P		6:05P	M	ML★ 1751	YBHQM	DO8	0
	6:35P		7:15P	O	UA★ 2856	YBMQH	F27	0
X6	6:40P		7:30P	O	AA★ 4189	YBMVQ	SH6	0
X6	8:25P		9:05P	O	UA★ 2858	YBMQH	F27	0

COMPREHEND
In the space below, write what you are to do.

_____

Write the steps you should follow to solve the problem.

_____

_____

_____

Write the information you need to solve the problem.

_____

_____

Write the system of measurement.

_____

COMPUTE
Do the computation required to complete each task.

_____

_____

_____

_____

COMMUNICATE
Write the flight numbers, times, and airlines as you would put them in a note to your supervisor.

_____

_____

_____

_____

_____

5 Your supervisor wants to fly to Missoula, Montana, after 4:30 p.m. Tuesday and return to Chicago around noon on Friday. Pick a flight. Write the airline, times, and flight number.

To MISSOULA, MONTANA					MDT	MSQ
▲CHICAGO, ILLINOIS					CDT	CHI
M-MDW O-ORD C-CGX P-PWK						
X6	7:00a O 11:45a	NW 707	FYBQM	72S	B/S	2
6	7:00a O 11:45a	NW 707	FYBQM	•	B/S	2
		NW 707 D9S-MSP-72S				
CONNECTIONS						
7:00a M	8:21a MSP	NW 611	FYBQM	72S	S	0
2AUG 9:10a MSP **11:55a**	NW 707	FYBQM	72S			1
8:00a O	10:09a SLC	DL 859	FYBMQ	757	B	0
11:16a SLC **12:50P**	DL 1829	FYBMQ	73S	S		0
11:50a O	1:59P SLC	DL 1489	FYBMQ	72S	L	0
2:59P SLC	**4:20P**	DL 1825	FYBMQ	73S		0
4:40P M	6:09P MSP	NW 507	FYBQM	727		0
7:00P MSP	9:25P	NW 709	FYBQM	D9S		1
8:00P O	6:25P MSP	NW 141	FYBQM	747	S	0
7:00P MSP	9:25P	NW 709	FYBQM	D9S		1

To CHICAGO, ILLINOIS					CDT	CHI
▲MISSOULA, MONTANA					MDT	MSQ
12:25P	7:03P M	NW 708	FYBQM	72S		2
CONNECTIONS	NW 707 D9S-MSP-72S					
7:00a	8:17a SLC	DL 1504	FYBMQ	73S	S	0
9:32a SLC **1:30P**	DL 614	FYBMQ	72S	B		0
8:00a	12:15P MSP	NW 706	FYBQM	D9S	S	1
1:00P MSP **2:15P**	NW 134	FYBMQ	M80	S		0
8:00a	12:15P MSP	NW 706	FYBQM	D9S	S	1
1:20P MSP **2:45P**	NW 676	FYBQM	727			0
12:25P	4:50P MSP	NW 708	FYBQM	72S		1
6:00P MSP **7:17P**	NW 784	FYBQM	72S	S		0
2:15P	3:29P SLC	DL 1826	FYBMQ	73S		0
4:17P SLC **8:10P**	DL 906	FYBMQ	72S	D		0
4:45P	7:08P SLC	DL 1825	FYBMQ	73S	S	1
7:51P SLC **11:40P**	DL 1488	FYBMQ	72S	S		0

Table 5-7

COMPREHEND
In the space below, write what you are to do.

_____

Write the steps you should follow to solve the problem.

_____

_____

_____

Write the information you need to solve the problem.

_____

_____

Write the system of measurement.

_____

COMPUTE
Do the computation required to complete each task.

_____
_____
_____
_____
_____
_____
_____
_____
_____

COMMUNICATE
Write the flight numbers, times, and airlines as you would put them in
a note to your supervisor.

_____
_____

TASK 2: CALCULATING CAR RENTAL COSTS

As an office assistant, you need to find car rental costs twice when you
are renting a car for your supervisor. The first time is when you are re-
serving a rental car for your supervisor. This gives you an estimate to use
in estimating the total cost of the trip. The second time is when you are
checking the bill from the car rental company.

To calculate the cost of a rental car, you'll use the car rental agency
rates shown in Figure 4-5 in the *Knowledge Base*.

Here's how you calculate the cost of a rental car.

1. Multiply the number of days times the correct daily rate (weekday
   or weekend).
2. Multiply the miles driven (or estimated to be driven) times the mileage fee.
3. Multiply the daily insurance cost times the number of days.
4. Add these three fees to get the total fee.

In addition, most states charge sales tax on a rental car. This will be
added to the total charge.

When the car rental company offers special rates, you always check
to see whether the special will be a better choice. To do this, you calculate the rate as usual. Then you compare your total to the total for the
special rate to see which is cheaper.

USING THE 3 CS

Your supervisor wants to rent a compact car from Monday to Friday. He plans to drive about 500 miles. Estimate the total rental costs. (Use the rates in Figure 4-5 in the *Knowledge Base*.)

COMPREHEND

In the space below, write what you are to do.

Estimate the total car rental costs.

Write the steps you should follow to solve the problem.

1. Multiply the number of days times the correct daily rate.
2. Multiply the miles times the mileage fee.
3. Multiply the daily insurance cost times the number of days.
4. Add these three fees to get the total fee.

Write the information you need to solve the problem.

Compact car, Monday to Friday (5 days), 500 miles

Write the system of measurement.

Money, time, and distance

COMPUTE

Do the computation required to complete each task.
Use the formula given above.

Daily rate 24.95×5 days = $124.75
Mileage rate 16¢ per mile $\times$ 500 miles = $80
Insurance rate $6.00 per day $\times$ 5 days = $30
Total: $124.75 + $80 + $30 = $234.75

COMMUNICATE

Write the estimate as you would write it in a budget for the trip.

$234.75

PRACTICE

6 Your supervisor wants to rent a luxury car for one weekday. She estimates she will drive 100 miles. Estimate the total car rental costs.

COMPREHEND

In the space below, write what you are to do.

_____

Write the steps you should follow to solve the problem.

_____

_____

_____

_____

Write the information you need to solve the problem.

_____

Write the system of measurement.

_____

COMPUTE
Do the computation required to complete each task.
Use the formula given on page 101.

_____

_____

_____

_____

COMMUNICATE
Write the estimate as you would write it in a budget for the trip.

_____

7 Your supervisor wants to rent a van for Tuesday and Wednesday. He estimates he will drive 700 miles. Estimate the total car rental costs.

COMPREHEND
In the space below, write what you are to do.

_____

Write the steps you should follow to solve the problem.

_____

_____

_____

_____

Write the information you need to solve the problem.

_____

Write the system of measurement.

_____

COMPUTE
Do the computation required to complete each task.
Use the formula given on page 101.

_____

_____

_____

_____

COMMUNICATE
Write the estimate as you would write it in a budget for the trip.

_____

8 Your supervisor wants to rent a four-door sedan for Friday and Saturday. She estimates she will drive 200 miles. Estimate the total car rental costs.

COMPREHEND
In the space below, write what you are to do.

_____

Write the steps you should follow to solve the problem.

_____

_____

_____

Write the information you need to solve the problem.

_____

_____

Write the system of measurement.

_____

COMPUTE
Do the computation required to complete each task.
Use the formula given on page 101.

_____

_____

_____

_____

_____

COMMUNICATE
Write the estimate as you would write it in a budget for the trip.

_____

9 Your supervisor wants to rent a subcompact for a week, Sunday to Saturday. He estimates he will drive 400 miles. Estimate the total car rental costs.

COMPREHEND
In the space below, write what you are to do.

_____

Write the steps you should follow to solve the problem.

_____

_____

_____

_____

Write the information you need to solve the problem.

_____

_____

Write the system of measurement.

_____

COMPUTE
Do the computation required to complete each task.
Use the formula given on page 101.

_____

_____

_____

_____

_____

COMMUNICATE
Write the estimate as you would write it in a budget for the trip.

_____

10 Your supervisor wants to rent a four-door sedan for one weekday and expects to drive 100 miles. The car rental company is offering a special on four-door sedan rentals for any weekday. The cost is $45, including mileage and insurance. Estimate the total car rental costs on the daily rate. Decide which is less expensive for this rental, the regular rate or the special rate.

COMPREHEND
In the space below, write what you are to do.

_____

_____

Write the steps you should follow to solve the problem.

_____

_____

_____

_____

_____

Write the information you need to solve the problem.

_____

Write the system of measurement.

_____

COMPUTE
Do the computation required to complete each task.
Use the formula given on page 101.

_____

_____

_____

_____

COMMUNICATE
Write which rate is cheaper. Write the estimate as you would write it in a budget for the trip.

_____

11 Your supervisor wants to rent a subcompact from Monday to Thursday and expects to drive 300 miles. The car rental company is offering a special on subcompact rentals for Monday through Friday. The cost is $150, including mileage and insurance. Estimate the total car rental costs on the daily rate. Decide which is less expensive for this rental, the regular rate or the special rate.

COMPREHEND
In the space below, write what you are to do.

_____

_____

Write the steps you should follow to solve the problem.

_____

_____

_____

_____

_____

Write the information you need to solve the problem.

_____

Write the system of measurement.

_____

COMPUTE
Do the computation required to complete each task.
Use the formula given on page 101.

_____

_____

_____

_____

COMMUNICATE
Write which rate is cheaper. Write the estimate as you would write it
in a budget for the trip.

_____

MATHEMATICS AND HUMAN RESOURCES

Part 6 covers some human resources you may deal with on the job. There are two job situations in this part of the book: Producing Graphs and Obtaining Data from a Database. The subject that you'll be dealing with in each job situation is people—employees of your company. Each job situation requires you to do several math tasks. Use the 3 Cs to complete each math task.

❏ JOB SITUATION 1 PRODUCING GRAPHS

As an office assistant at Gibson Paper Company, you often have to include graphs in the reports you prepare. You use different methods to produce the graphs. When the graph is for use only by Gibson's own employees, it may not have to look perfect. You may draw the graph yourself. You'll use graph paper and draw and label the graph by hand. Sometimes, for this kind of graph, you'll use some computer tools. There are software packages that allow you to tell the computer to draw a graph. They produce a graph that can be printed on one of the company's printers.

When the graph is going to Gibson's customers, it has to look better. You'll probably use computer tools to produce this kind of graph.

Sometimes, you are asked to prepare a report that will be given to the Desktop Publishing Department to lay out and print. This might be a sales brochure or the company's annual report. The graphs in this kind of report will be done by a desktop publishing specialist. You will need to supply a sketch of the graph. Again, you will use graph paper to draw the sketch yourself.

Whichever method you use to produce the graph, you have to make the same choices about it. You need to know what type of graph to use. You have a choice of three main types of graphs: bar graphs, line graphs, and circle graphs. (Circle graphs are also called pie charts.) Figures 5-10, 5-11, and 5-12 in the *Knowledge Base* show these graphs. Each one is good for showing a particular kind of information.

Circle graphs allow you to compare parts of a whole. The circle graph in the *Knowledge Base* shows how much Gibson's salespeople have sold in different states. The whole circle stands for Gibson's total sales. Each wedge of the circle stands for sales to a different state.

Bar graphs are good for comparing different numbers. The bar graph in Figure 5-11 shows the total the sales force has sold to each of four states. From the graph you can get the dollar figure for sales to each state. You can also see which state has the greatest sales.

Line graphs are used to show numbers and also to show changes over time. The line graph in Figure 5-12 shows sales over a period of four years. You can see what the sales were each year. You can also see when sales went up or down.

Once you have picked the type of graph to draw, you need to make some other decisions. You have to decide what labels to give the graph and what scale to use. You also decide how to draw the lines, wedges, or bars. Sometimes you will go ahead and draw the graph by hand. Other times, you will enter these choices into the computer and the computer will produce the graph.

TASK 1: MAKING A BAR GRAPH

When you decide to use a bar graph, you should begin by preparing a table of the information you want to show. This will help you to draw the graph. Suppose you had drawn the bar graph in Figure 5-11 of the *Knowledge Base*. First you drew this table to show the information.

State	Total Sales
IA	$68,000
IN	$68,000
IL	$750,000
MO	$477,000

Next, you chose a label for each **axis** of the graph. The **axes** of a graph are the vertical and horizontal lines. One axis will show the states. One will show the sales figures. You chose to label the vertical axis "States." You labeled the horizontal axis "Sales in Thousands of Dollars." Showing the dollars as thousands of dollars means you don't have to put extra zeros for all the dollar amounts.

You wrote each item on each scale. On the vertical axis, you wrote the name of each state. Then you decided on the scale for the numerical axis. (A scale should include the lowest to the highest number you need to show. It can go from just below to just above these numbers. The interval should be small enough to show the differences between the data but large enough so that you don't crowd the scale with extra numbers.) You chose to have a scale going from $0 to $800,000. The interval between labels was $100,000. So the scale reads $100,000, $200,000, $300,000, and so on.

Next you drew the bars to show each number in your table. For example, the bar for Iowa goes to a point between $50,000 and $100,000 on the scale in order to show $68,000.

Finally, you wrote a title for the graph—"1992 Sales."

These are the steps you follow in creating any bar graph. To sum up:

1. Create a table for the data.
2. Choose a label for each axis.

3. Write the data on each axis. Choose a scale for the numerical axis.

4. Draw a bar for each pair of numbers in your table.

5. Title the graph.

Remember, these steps apply whether you draw the graph by hand or enter the information into a computer graphing tool.

USING THE 3 Cs

Draw a bar graph to show the total sales for each of these sales people: Brown, Jones, Smith, and Anderson. Brown sold $100,000 in 1992. Jones sold $90,000. Smith sold $50,000. Anderson sold $120,000.

COMPREHEND

In the space below, write what you are to do.

Draw a bar graph to show sales per salesperson.

Write the steps you should follow to solve the problem.

1. Create a table for the data.

2. Choose a label for each axis.

3. Write the data on each axis. Choose a scale and interval for the numerical axis.

4. Draw a bar for each pair of numbers in your table.

5. Title the graph.

Write the information you need to solve the problem.

Brown sold $100,000 in 1992. Jones sold $90,000. Smith sold $50,000. Anderson sold $120,000.

Write the system of measurement.

Money

COMPUTE

Do the computation required to complete each task. Use the instructions given above.

1. Table:

Salesperson	Sale
Brown	$100,000
Jones	$90,000
Smith	$50,000
Anderson	$120,000

2. Labels: "Salesperson"; "Sales in Thousands of Dollars"

3. Scale: $0 to $150,000; Interval: $10,000

4. Bars: See the graph on page 110.

5. Title: "1992 Sales"

COMMUNICATE

Draw the graph. Use the graph paper shown here.

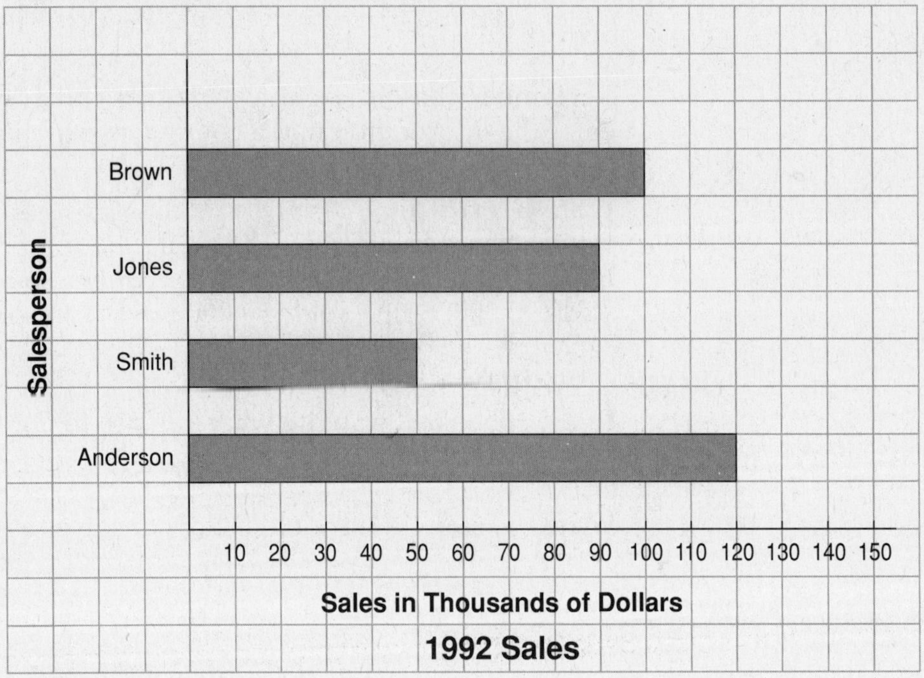
Brown
Jones
Smith
Anderson
Salesperson
10 20 30 40 50 60 70 80 90 100 110 120 130 140 150
Sales in Thousands of Dollars
1992 Sales

PRACTICE

1 Draw a bar graph to show the total sales for each of these sales people: Marcus, Williams, and Macintosh. Marcus sold $80,000 in 1992. Williams sold $40,000. Macintosh sold $90,000.

COMPREHEND

In the space below, write what you are to do.

____

Write the steps you should follow to solve the problem.

____

____

____

____

____

____

Write the information you need to solve the problem.

____

____

Write the system of measurement.

____

COMPUTE

Do the computation required to complete each task.

Use the instructions given on page 109.

_____
_____
_____
_____

COMMUNICATE
Draw the graph. Use the graph paper shown here.

2 Draw a bar graph to show the total sales for each of these sales groups: Group 1, Group 2, and Group 3. Group 1 sold $200,000 in 1991. Group 2 sold $100,000. Group 3 sold $150,000.

COMPREHEND
In the space below, write what you are to do.

_____

Write the steps you should follow to solve the problem.

_____
_____
_____
_____
_____
_____

Write the information you need to solve the problem.

_____

_____

Write the system of measurement.

_____

COMPUTE
Do the computation required to complete each task.
Use the instructions given on page 109.

_____

_____

_____

_____

_____

COMMUNICATE
Draw the graph. Use the graph paper shown here.

3 Draw a graph to show the total sales for each of these sales people: Saunders, Miller, White, and Mendez. Saunders sold $20,000 in July. Miller sold $15,000 in July. White sold $10,000 in July. Mendez sold $20,000 in July.

COMPREHEND
In the space below, write what you are to do.

_____

Write the steps you should follow to solve the problem.

_____

_____

_____

_____

_____

_____

Write the information you need to solve the problem.

_____

_____

_____

Write the system of measurement.

_____

COMPUTE
Do the computation required to complete each task.
Use the instructions given on page 109.

_____

_____

_____

_____

_____

COMMUNICATE
Draw the graph. Use the graph paper shown here.

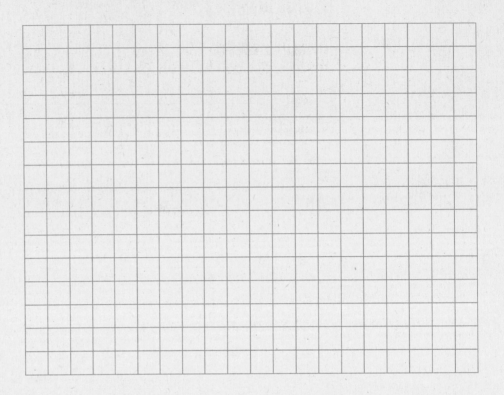

4 Draw a bar graph to show the total sales for each of these four years: 1989, 1990, 1991, and 1992. The sales were $1,000,000 in 1989; $1,100,000 in 1990; $1,200,000 in 1991; and $1,350,000 in 1992.

COMPREHEND
In the space below, write what you are to do.

_____

Write the steps you should follow to solve the problem.

_____

_____

_____

_____

_____

Write the information you need to solve the problem.

_____

_____

Write the system of measurement.

_____

COMPUTE
Do the computation required to complete each task.
Use the instructions given on page 109.

_____

_____

_____

_____

_____

COMMUNICATE
Draw the graph. Use the graph paper shown here.

TASK 2: CREATING A LINE GRAPH

When you decide to use a line graph to show information, you follow steps very similar to those you use in making a bar graph.

First make a table showing the information. Here is the table that you would have drawn in making the line graph shown in Figure 5-12 of the *Knowledge Base*.

Year	Total Sales
1988	$100,000
1989	$300,000
1990	$275,000
1991	$477,000
1992	$750,000

Next, you chose a label for each axis of the graph.

You wrote each item on each scale. On the vertical axis, you wrote each year. Then you decided on the scale for the numerical axis. You decided to have a scale going from $100,000 to $800,000. The interval between labels was $100,000.

Next you drew a point on the graph for every sales figure in the table. You connected the points with lines.

Finally, you wrote a title for the graph—"Sales in Illinois."

These are the steps you follow in creating any line graph. To sum up:

1. Create a table for the data.

2. Choose a label for each axis.

3. Write the data on each axis. Choose a scale for the numerical axis.

4. Draw a point for each number in your table. Connect the points with lines.

5. Title the graph.

Remember, these steps apply whether you draw the graph by hand or enter the information into a computer.

USING THE 3 CS

Draw a line graph to show Brown's sales from 1989 to 1992. Sales were $75,000 in 1989; $80,000 in 1990; $100,000 in 1991; and $100,000 in 1992.

COMPREHEND

In the space below, write what you are to do.

Draw a line graph to show Brown's sales each year.

Write the steps you should follow to solve the problem.

1. Create a table for the data.

2. Choose a label for each axis.

3. Write the data on each axis. Choose a scale and interval for the numerical axis.

4. Draw a point for each number. Connect the points with lines.
5. Title the graph.

Write the information you need to solve the problem.

Sales were $75,000 in 1989; $80,000 in 1990; $100,000 in 1991; $100,000 in 1992.

Write the system of measurement.

Money

COMPUTE
Do the computation required to complete each task.
Use the instructions given on page 116.

1. Table:

Year	Sales
1989	$75,000
1990	$80,000
1991	$100,000
1992	$100,000

2. Labels: "Years"; "Sales in Thousands of Dollars"
3. Scale: $75,000 to $110,000; Interval: $10,000
4. Lines: See the graph below.
5. Title: "Brown's Sales"

COMMUNICATE
Draw the graph. Use the graph paper shown here.

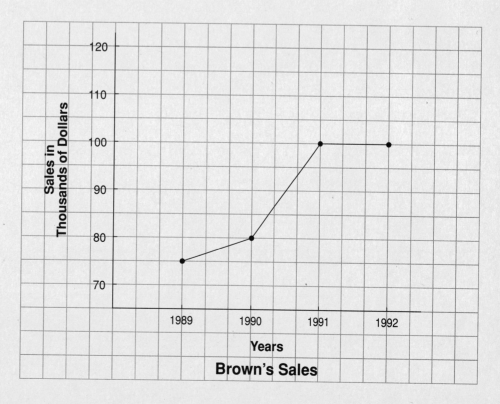

5 | Draw a line graph to show Anderson's sales from 1989 to 1992. Sales were $75,000 in 1989; $90,000 in 1990; $100,000 in 1991; and $120,000 in 1992.

COMPREHEND

In the space below, write what you are to do.

_____

Write the steps you should follow to solve the problem.

_____

_____

_____

_____

_____

_____

Write the information you need to solve the problem.

_____

_____

Write the system of measurement.

_____

COMPUTE

Do the computation required to complete each task.
Use the instructions given on page 116.

_____

_____

_____

_____

_____

COMMUNICATE
Draw the graph. Use the graph paper shown here.

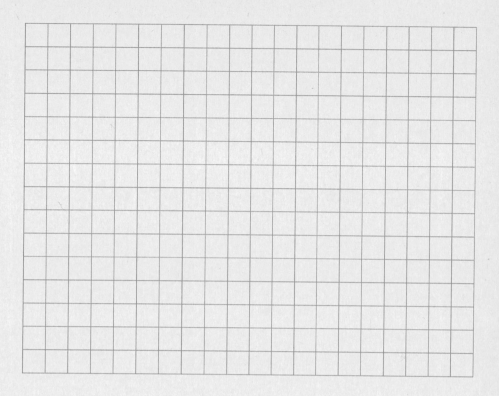

6 Draw a line graph to show Group 1's sales from 1989 to 1992. Sales were $135,000 in 1989; $145,000 in 1990; $180,000 in 1991; and $200,000 in 1992.

COMPREHEND
In the space below, write what you are to do.

_____

Write the steps you should follow to solve the problem.

_____

_____

_____

_____

_____

_____

Write the information you need to solve the problem.

_____

_____

Write the system of measurement.

_____

COMPUTE
Do the computation required to complete each task.
Use the instructions given on page 116.

_____

_____

_____

_____

_____

COMMUNICATE
Draw the graph. Use the graph paper shown here.

7 Draw a line graph to show sales in Illinois for the months of January through May. Sales were $50,000 in January, $34,500 in February; $40,000 in March; $60,000 in April; and $62,000 in May.

COMPREHEND
In the space below, write what you are to do.

_____

_____

Write the steps you should follow to solve the problem.

_____

_____

_____

_____

_____

_____

_____

Write the information you need to solve the problem.

_____

_____

Write the system of measurement.

_____

COMPUTE
Do the computation required to complete each task.
Use the instructions given on page 116.

_____

_____

_____

_____

_____

COMMUNICATE
Draw the graph. Use the graph paper shown here.

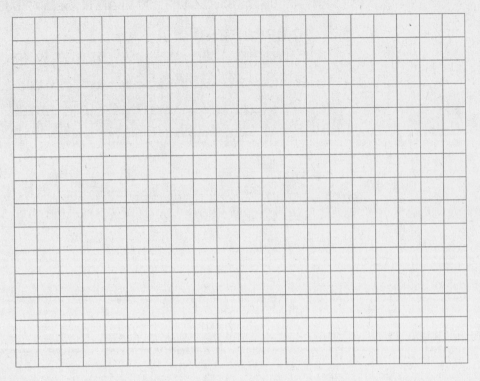

8 Draw a line graph to show sales in Iowa for the months of June through October. Sales were $6,000 in June; $4,500 in July; $4,000 in August; $7,100 in September; and $7,000 in October.

COMPREHEND
In the space below, write what you are to do.

_____

_____

Write the steps you should follow to solve the problem.

_____

_____

_____

_____

_____

_____

Write the information you need to solve the problem.

_____

_____

_____

Write the system of measurement.

_____

COMPUTE
Do the computation required to complete each task.
Use the instructions given on page 116.

_____

_____

_____

_____

COMMUNICATE
Draw the graph. Use the graph paper shown here.

TASK 3: CREATING A CIRCLE GRAPH

Circle graphs show the relationships between parts of a whole. The circle stands for the whole amount. The whole may be the total money spent in a month, as in the circle graph, or pie chart, in Figure 5-10 of the *Knowledge Base*. Each section, or wedge, of the circle stands for one of the parts of the whole. Each wedge in Figure 5-10 is one of the states where sales are made. Each wedge is a percentage of the whole. The whole is 100 percent. The larger the percentage of the whole, the larger the wedge.

Here are the steps you would have followed to create the circle graph shown in Figure 5-10.

First you made a table showing the information. The table shows the value of each part of the whole. Here is the table that you drew for the circle graph in Figure 5-10.

State	IL	IA	MO	IN
Sales	750,000	68,000	477,000	68,000

Next you found the percentage of the whole that each item stands for.

Illinois: 55%   Iowa: 5%   Missouri: 35%   Indiana: 5%

Then you found the number of degrees that each percentage stands for. (There are 360° in a circle.)

Illinois: 198°   Iowa: 18°   Missouri: 126°   Indiana: 18°

Then you drew the circle. You used a **protractor**, a tool that measures angels, to draw the angle (wedge) for each state. You labeled the angle.

Finally, you wrote a title for the graph.

These are the steps you follow in creating any circle graph. To sum up:

1. Create a table for the data.

2. Find the percentage of the whole for each item.

3. Find the number of degrees for each item.

4. Draw a circle and draw an angle for each item of data. Label it.

5. Title the graph.

Remember, these steps apply whether you draw the graph by hand or enter the information into a computer.

USING THE 3 CS

Employees in the Data Processing Department missed a total of 500 days of last year. Draw a circle graph to show the causes for the days not worked. The causes were 250 vacation days; 175 sick days; 50 personal days; and 25 jury duty days.

COMPREHEND

In the space below, write what you are to do.

Make a circle graph to show days missed from work in the Data Processing Department.

Write the steps you should follow to solve the problem.

1. Create a table for the data.
2. Find the percentage of the whole for each item.
3. Find the number of degrees for each item.
4. Draw a circle and draw an angle for each item of data. Label it.
5. Title the graph.

Write the information you need to solve the problem.

Total of 500 days not worked last year. The causes were 250 vacation days; 175 sick days; 50 personal days; and 25 jury duty days.

Write the system of measurement.

Time

COMPUTE
Do the computation required to complete each task.
Use the instructions given on page 124.

1. Table:

Cause	Vacation	Sick	Personal	Jury Duty
Days	250	175	50	25

2. Percentages:

Vacation days 250 out of 500. What percent is 250 of 500?
$n \times 500 \quad = 250$
$\div 500 \qquad \div 500$ (Divide both sides of the equation by 500.)
$n = 0.50 \quad = 50\%$

Sick days 175 out of 500. What percent is 175 of 500?
$n \times 500 \quad = 175$
$n = 0.35 \quad = 35\%$

Personal days 50 out of 500. What percent is 50 of 500?
$n \times 500 \quad = 50$
$n = 0.1 \quad = 10\%$

Jury Duty days 25 out of 500. What percent is 25 of 500?
$n \times 500 \quad = 25$
$n = 0.05 \quad = 5\%$

3. Angles:

Vacation days 50%. Angle is 50% of 360°.
$360° \times 0.50 = 180°$

Sick days 35%. Angle is 35% of 360°.
$360° \times 0.35 = 126°$

Personal days 10%.
$360° \times 0.10 = 36°$

Jury duty 5%.
$360° \times 0.05 = 18°$

4. Draw a circle with a compass.

Use a protractor to draw each angle.

- Draw a line from the center of the circle to one side.

- *Put the center of the protractor on the center of the circle and the 0° mark on the protractor on the line.*

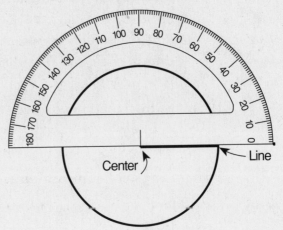

- *To draw an angle of 36°, put a dot on the paper at the 36° mark on the protractor. Line up a ruler from the dot at the center of the circle to the dot. Draw a line from the center of the circle to the edge of the circle.*

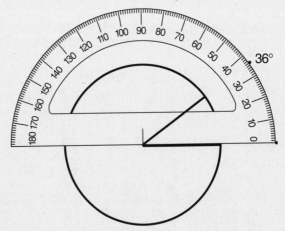

- *To draw the next angle, place the center of the protractor at the center of the circle, but this time put the 0° mark on the line you just drew. (Turn the paper as necessary.)*

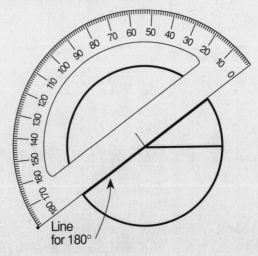

Be sure to label each angle.

5. *Title: "Days Not Worked in Data Processing Department"*

Part 6: Mathematics and Human Resources

COMMUNICATE
Make the graph.

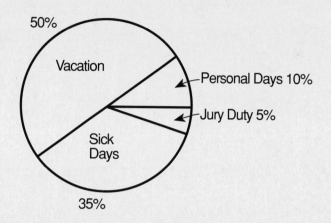

Days Not Worked in Data Processing Department

PRACTICE

9 Draw a circle graph to show days not worked in the Accounting Department. Total days: 100. Causes: 50 vacation days; 25 sick days; 25 personal days.

COMPREHEND
In the space below, write what you are to do.

_____

_____

Write the steps you should follow to solve the problem.

_____

_____

_____

_____

_____

Write the information you need to solve the problem.

_____

_____

Write the system of measurement.

_____

COMPUTE
Do the computation required to complete each task.
Use the instructions given on page 124.

Cause	Vacation	Sick	Personal
Days			

_____

_____

_____

COMMUNICATE
Draw the graph.

[10] Draw a circle graph to show days not worked in the Mailroom. Total days: 200. Causes: 100 vacation days; 60 sick days; 40 personal days.

COMPREHEND
In the space below, write what you are to do.

_____

_____

Write the steps you should follow to solve the problem.

_____

_____

_____

_____

_____

_____

Write the information you need to solve the problem.

_____

_____

Write the system of measurement.

_____

COMPUTE
Do the computation required to complete each task.
Use the instructions given on page 124.

Cause	Vacation	Sick	Personal
Days			

_____

_____

_____

COMMUNICATE
Draw the graph.

11 Draw a circle graph to show the number of employees in different departments. Total employees: 200. Departments: data processing—50; accounting—20; mailroom—10; sales—70; office assistants—50.

COMPREHEND
In the space below, write what you are to do.

_____

_____

Write the steps you should follow to solve the problem.

_____

_____

_____

_____

_____

_____

Write the information you need to solve the problem.

_____

_____

_____

Write the system of measurement.

_____

COMPUTE
Do the computation required to complete each task.
Use the instructions given on page 124.

_____

_____

_____

_____

_____

COMMUNICATE
Draw the graph.

12 Draw a circle graph to show the number of salespeople in each state. Total salespeople: 150. Illinois—65; Iowa—25; Missouri—45; Indiana—15.

COMPREHEND

In the space below, write what you are to do.

_____

_____

Write the steps you should follow to solve the problem.

_____

_____

_____

_____

_____

_____

Write the information you need to solve the problem.

_____

_____

Write the system of measurement.

_____

_____

COMPUTE

Do the computation required to complete each task.
Use the instructions given on page 124.

_____

_____

_____

_____

COMMUNICATE

Draw the graph.

❏ JOB SITUATION 2 OBTAINING DATA FROM A DATABASE

When you work in the Data Processing Department at Gibson Paper Company, you work with a number of different databases. Chapter 6 of the *Office Technology Knowledge Base* discusses databases. One day you work on a database of Gibson's customers. Another day you work on a database of Gibson's products. Sometimes you'll be entering or changing data in the database. At other times, you'll be asked to get information from the database. You will **select** the right data from the database, **sort** it, and **print** the information.

For example, you are asked to get the list of current customers and their addresses from the database. You are asked to have the report in alphabetical order. You work on the database called "Mail," for "Mailing List." You give the database program the command to **select** current customers. Part of what you see on the screen is shown in Figure 6-9 of the *Knowledge Base*. Then you give the command to **sort** the information alphabetically. The result looks like Figure 6-12 of the *Knowledge Base*. Finally, you **print** the report, making sure it includes the addresses.

Sometimes you are also asked to get statistics on data from the database. For example, you are asked to find the mean yearly sales to a customer for the last five years. Or you are asked to find the median level of monthly sales in Illinois for the year.

Some database programs may be set up so that there are commands to calculate these statistics. At other times, however, you will need to:

❏ First obtain the information you need from the database. For example, to find the mean sales to a customer for the last five years, get a report containing total sales to this customer for each of the five years.

❏ Then do the math to get the statistics you need.

Even if you can have the computer calculate a statistic, you should always check to see if the answer seems reasonable. If you enter the wrong commands by mistake, the calculation will be wrong.

TASK 1: CALCULATING MEAN WAGES

You are asked to use the company's payroll database. You will calculate the mean wage paid to different groups of employees. Part of the payroll database is shown in Figure 6-10 of the *Knowledge Base*. Each line in the database is a **record** of information about a different employee. Each item in the line is one **field**, containing a specific piece of information. Figure 6-11 of the *Knowledge Base* shows what each field stands for.

Once you have selected, sorted, and printed the data, you are ready to calculate the mean wage. You follow these steps:

1. Determine which field or fields contain the information you need.

2. Copy the necessary numbers.

3. Calculate the mean, using this formula:

$$\text{mean} = (\text{total of the numbers}) \div (\text{number of items})$$

USING THE 3 CS

You have a request from the Accounting Department to find the mean hourly wage paid to the employees shown in the database in Figure 6-11 of the Knowledge Base.

COMPREHEND

In the space below, write what you are to do.

Find the mean hourly wage for the employees in the database.

Write the steps you should follow to solve the problem.

1. Determine which field or fields contain the information.
2. Copy the necessary numbers.
3. Calculate the mean.

Write the information you need to solve the problem.

Information is found in the field labeled "Wage."
Wages: $8.90 $8.45 $6.70 $8.45 $8.40

Write the system of measurement.

Money

COMPUTE

Do the computation required to complete each task.
Use the formula given above.

mean = (total of the numbers) ÷ (number of items)
mean = ($8.90 + $8.45 + $6.70 + $8.45 + $8.40) ÷ 5
mean = $40.90 ÷ 5
mean = $8.18

COMMUNICATE

Write the answer to the request.
The mean hourly wage for these employees is $8.18.

PRACTICE

1 You have a request to find the mean overtime wage for the employees in the group shown in Figure 6-11 of the *Knowledge Base*. Do the necessary calculations.

COMPREHEND

In the space below, write what you are to do.

_____

_____

Write the steps you should follow to solve the problem.

_____

_____

_____

_____

Write the information you need to solve the problem.

_____

_____

Write the system of measurement.

_____

COMPUTE
Do the computation required to complete each task.
Use the formula given on page 133.

COMMUNICATE
Write the answer to the request.

_____

_____

2   You have a request to find the mean hourly wage for four employees in the Data Processing Department. You have selected the employees from the database and printed this report.

```
Record      Name        SS Number     Wage  Hours
  1       Lopez     B  334778432      8.45   40
  2       Pitini    G  555843222     12.45   40
  3       Andrews   A  326778107      7.90   35
  4       Brown     G  121456887      8.25   40
```

Do the necessary calculations to find the mean wage.

COMPREHEND
In the space below, write what you are to do.

_____

_____

Write the steps you should follow to solve the problem.

_____

_____

_____

_____

Write the information you need to solve the problem.

_____

_____

Write the system of measurement.

_____

COMPUTE
Do the computation required to complete each task.
Use the formula given on page 133.

COMMUNICATE
Write the answer to the request.

_____

3 You have a request to find the mean hourly wage for five employees in the Mailroom. You have selected the correct employees from the database and printed this report.

Record	Name		SS Number	Wage	Hours
1	Chu	B	475445667	6.70	40
2	Masillo	C	782334667	5.75	40
3	Wilson	J	115938372	4.95	40
4	Cronin	A	893220122	5.00	40
5	Manjarez	M	017225346	6.25	40

Do the necessary calculations to find the mean wage.

COMPREHEND
In the space below, write what you are to do.

_____

_____

Write the steps you should follow to solve the problem.

_____

_____

_____

_____

Write the information you need to solve the problem.

_____

_____

Write the system of measurement.

_____

COMPUTE
Do the computation required to complete each task.
Use the formula given on page 133.

COMMUNICATE
Write the answer to the request.

_____

_____

4 You have a request to find the mean hourly wage for these six employees in the Copy Center. You have selected the employees from the database and printed this report.

```
Record    Name            SS Number    Wage  Hours
  1       Bayard    M  126348755      8.90   40
  2       Ross      C  475678111      7.75   40
  3       Washington J 023455619      6.50   40
  4       Morris    C  234578112      5.90   40
  5       Ravi      M  789223586      8.00   40
  6       Peterson  S  873102233      7.50   40
```

Do the necessary calculations to find the mean wage.

COMPREHEND
In the space below, write what you are to do.

_____

_____

Write the steps you should follow to solve the problem.

_____

_____

_____

_____

Write the information you need to solve the problem.

_____

_____

Write the system of measurement.

_____

COMPUTE
Do the computation required to complete each task.
Use the formula given on page 133.

COMMUNICATE
Write the answer to the request.

_____

_____

5 You have a request to find the mean hourly wage for a group of six office assistants. You have selected the employees from the database and printed this report.

Record	Name		SS Number	Wage	Hours
1	Stein	J	136877643	9.60	40
2	Lopez	J	783112353	8.75	40
3	Wu	B	120938750	9.00	40
4	Howard	A	134375672	8.00	40
5	Kooper	S	359629522	9.00	40
6	Krasnow	M	679132239	8.25	40

Do the necessary calculations to find the mean wage.

COMPREHEND
In the space below, write what you are to do.

_____

_____

Write the steps you should follow to solve the problem.

_____

_____

_____

_____

Write the information you need to solve the problem.

_____

_____

Write the system of measurement.

_____

COMPUTE
Do the computation required to complete each task.
Use the formula given on page 133.

COMMUNICATE
Write the answer to the request.

_____

_____

TASK 2: CALCULATING OTHER STATISTICS ABOUT EMPLOYEES

Gibson uses a database to keep track of how many people are employed by each department of the company. Each department fills out a form telling the total of permanent and temporary employees in the department every month. You enter that information in the database. This is what a record in the database looks like when you print it:

Department	Jan	Feb	Mar	Apr	May	June	July	Aug	Sept	Oct	Nov	Dec
Desktop Pub.	20	19	20	20	21	17	15	15	16	19	20	20

Each field tells you how many employees were in the department that month.

You may be asked to calculate descriptive statistics, such as the range, mode, and median, from this data.

Once you have selected, sorted, and printed the data, you are ready to calculate the mean. You follow these steps:

1. Determine which field or fields contain the information you need.
2. Copy the necessary numbers, arranging them in order from least to greatest.
3. Determine the range, using this formula:

$$\text{range} = \text{greatest number} - \text{least number}$$

Determine the mode, using this rule:

Mode is the number that occurs most often in the group of data.

Determine the median, using this formula:

For an odd number of items, find the number in the middle.

For an even number of items, find the number just above and the number just below the middle. Find the mean of these two numbers.

USING THE 3 CS

You have a request from the Personnel Department for these statistics on the number of employees in the Desktop Publishing Department each month: range, mode, and median number of employees. Using the portion of the database shown on page 138, calculate these statistics.

COMPREHEND

In the space below, write what you are to do.

Find the range, mode, and median number of employees in the Desktop Publishing Department.

Write the steps you should follow to solve the problem.

1. Determine which field or fields contain the information you need.
2. Copy the necessary numbers, arranging them in order from least to greatest.
3. Determine the range, mode, and median.

Write the information you need to solve the problem.

Information is found in the fields labeled by months. Arranged in order: 15 15 16 17 19 19 20 20 20 20 20 21

Write the system of measurement.
People

COMPUTE

Do the computation required to complete each task.
Use the formulas given above.

range: 21 - 15 = 6

total of 5

mode: 15 15 16 17 19 19 20 20 20 20 20 21
The most common is 20.

before   after
middle   middle
↓   ↓
median: 15 15 16 17 19 19 20 20 20 20 20 21
19 + 20 = 39 ÷ 2 = 19.5

COMMUNICATE
Write the answer to the request.

The range is 6 employees. The mode for the number of employees is 20. The median number of employees is 19.5.

PRACTICE

6 You have a request for these statistics on the number of employees each month in the Mailroom: range, mode, and median number of employees. You have printed this report. Calculate the statistics.

Department	Jan	Feb	Mar	Apr	May	June	July	Aug	Sept	Oct	Nov	Dec
Mailroom	4	4	4	4	3	3	3	3	3	4	4	4

COMPREHEND
In the space below, write what you are to do.

_____

_____

Write the steps you should follow to solve the problem.

_____

_____

_____

_____

_____

Write the information you need to solve the problem.

_____

_____

Write the system of measurement.

_____

COMPUTE
Do the computation required to complete each task.
Use the formulas given on page 139.

_____

_____

_____

_____

_____

COMMUNICATE
Write the answer to the request.

_____

_____

7 You have a request for statistics on the number of employees each month in the Data Processing Department. Find the range, mode, and median number of employees. You have printed this report. Calculate the statistics.

Department	Jan	Feb	Mar	Apr	May	June	July	Aug	Sept	Oct	Nov	Dec
DP	8	8	9	6	7	7	7	7	9	9	7	9

COMPREHEND
In the space below, write what you are to do.

_____

_____

Write the steps you should follow to solve the problem.

_____

_____

_____

_____

Write the information you need to solve the problem.

_____

_____

Write the system of measurement.

_____

COMPUTE
Do the computation required to complete each task.
Use the formulas given on page 139.

_____

_____

_____

COMMUNICATE
Write the answer to the request.

_____

_____

8 You have a request for these statistics on the number of employees each month in the Sales Department: range, mode, and median number of employees. You have printed this report. Calculate the statistics.

Department	Jan	Feb	Mar	Apr	May	June	July	Aug	Sept	Oct	Nov	Dec
Sales	25	25	25	22	22	22	22	24	26	24	26	26

COMPREHEND
In the space below, write what you are to do.

_____

_____

Write the steps you should follow to solve the problem.

_____

_____

_____

_____

Write the information you need to solve the problem.

_____

_____

Write the system of measurement.

_____

COMPUTE
Do the computation required to complete each task.
Use the formulas given on page 139.

_____

_____

_____

COMMUNICATE
Write the answer to the request.

_____

_____

9 You have a request for these statistics on the number of employees each month in the Accounting Department: range, mode, and median number of employees. You have printed this report. Calculate the statistics.

Department	Jan	Feb	Mar	Apr	May	June	July	Aug	Sept	Oct	Nov	Dec
Accounting	5	5	12	13	10	9	7	5	10	11	6	7

COMPREHEND

In the space below, write what you are to do.

_____

_____

Write the steps you should follow to solve the problem.

_____

_____

_____

_____

_____

Write the information you need to solve the problem.

_____

_____

Write the system of measurement.

_____

COMPUTE

Do the computation required to complete each task.
Use the formulas given on page 139.

_____

_____

_____

COMMUNICATE

Write the answer to the request.

_____

_____

MATHEMATICS REVIEW

P A R T 7

❏ BASIC OPERATIONS

PLACE VALUE

The **digits** 0 to 9 are used to name any number. The **value** of a digit depends on its place in the number. For example, in each of these numbers, the digit 5 has a different value.

5	This 5 means 5 ones.
50	This 5 means 5 tens.
500	This 5 means 5 hundreds.
5,000	This 5 means 5 thousands.

To read any number, you need to know the value of each place. This chart shows **place value** up to hundred millions.

hundred millions	ten millions	millions	hundred thousands	ten thousands	thousands	hundreds	tens	ones

856,620,572

This number means:
8 hundred millions + 5 ten millions + 6 millions +
6 hundred thousands + 2 ten thousands + 0 thousands +
5 hundreds   + 7 tens   + 2 ones

78,019

This means:
7 ten thousands, 8 thousands
0 hundreds, 1 ten, 9 ones

Long numbers are separated into sections by commas. To read a number, read the value of each of the sections.

(32),(621)

Read as: thirty-two thousand, six hundred twenty-one

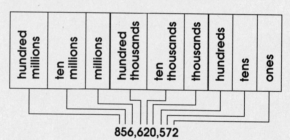

Read as:
eight hundred fifty-six million, six hundred twenty thousand, five hundred seventy-two

78,019

Read as: seventy-eight thousand, nineteen

150,399

Read as: one hundred fifty thousand, three hundred ninety-nine

15,399

Read as: fifteen thousand, three hundred ninety-nine

ADDITION

When adding any set of numbers, first write the numbers with the ones in the ones column, the tens in the tens column, and so on.

$$\text{Example:} \quad 912 + 36 \longrightarrow \quad \begin{array}{r} 9\ 1\ 2 \\ +\quad 3\ 6 \\ \hline \end{array}$$

Then follow these steps:

Step 1: Add the ones and write the answer in the ones column.

$$2 + 6 = 8$$

$$\begin{array}{r} 912 \\ +\ 36 \\ \hline 8 \end{array}$$

Step 2: Add the tens and write the answer in the tens column.

$$1 + 3 = 4$$

$$\begin{array}{r} 912 \\ +\ 36 \\ \hline 48 \end{array}$$

Step 3: Add the hundreds. Notice that in this case there is only one number in the hundreds column.

$$9 + 0 = 9$$

$$\begin{array}{r} 912 \\ +\ 36 \\ \hline 948 \end{array}$$

Write the 9 in the hundreds column. The answer is 948.

Here's an example that shows what to do when the numbers you get are too large for the column. See how you **regroup** when the number is too large.

$$647 + 598 \longrightarrow \quad \begin{array}{r} 647 \\ +598 \\ \hline \end{array}$$

Step 1: Add the ones.

$$7 + 8 = 15$$

You can't write 15 in the ones column, so think of 15 as 1 ten and 5 ones. Write the 5 ones in the ones column. Write the 1 ten at the top of the tens column.

$$\begin{array}{r} {\scriptstyle 1} \\ 647 \\ +598 \\ \hline 5 \end{array}$$

Step 2: Add the tens. Don't forget the 1 you wrote at the top of the tens column.

$1 + 4 + 9 = 14$

You can't write 14 in the tens column, so think of the 14 as 1 hundred and 4 tens. Write the 4 in the tens column and write the 1 hundred in the hundreds column.

$$\begin{array}{r} 1\,1 \\ 647 \\ +598 \\ \hline 45 \end{array}$$

Step 3: Add the hundreds.

$1 + 6 + 5 = 12$

Write the 2 in the hundreds column and write the 1 in the thousands column. The answer is 1,245.

$$\begin{array}{r} 1\,1 \\ 647 \\ +598 \\ \hline 1,245 \end{array}$$

To add larger numbers, follow the same steps. Write the number with the ones lined up, the tens lined up, and so on. Then add the ones, the tens, the hundreds, the thousands, and so on. Remember to write any number you have to regroup in the next column.

$$\begin{array}{r} 1 \\ 4,573 \\ +6,118 \\ \hline 10,691 \end{array} \qquad \begin{array}{r} 1\,2\,1 \\ 34,921 \\ 15,780 \\ 5,711 \\ +23,002 \\ \hline 79,414 \end{array} \qquad \begin{array}{r} 1\,2\,2 \\ 126,781 \\ 340,990 \\ +\;\;\;7,631 \\ \hline 475,402 \end{array} \qquad \begin{array}{r} 1\,1\,1\quad\;1 \\ 16,782,229 \\ +\;5,757,001 \\ \hline 22,539,230 \end{array}$$

SUBTRACTION

The **inverse** of addition is subtraction. Subtraction undoes addition. Every addition fact can be undone in two ways as subtraction facts.

Addition: $7 + 2 = 9$
Subtraction: $9 - 2 = 7$
                $9 - 7 = 2$

You can use what you know about addition to help you solve subtraction problems. Here's an example:

$$15 - 7 = ?$$

Ask "What number added to 7 gives 15?" The answer is 8.

$$15 - 7 = 8$$

Addition can also be used to check the answer in subtraction.

$$\begin{array}{r} 23 \\ -\;7 \\ \hline 16 \end{array} \qquad \text{Check:} \qquad \begin{array}{r} 16 \\ +\;7 \\ \hline 23\;\checkmark \end{array}$$

When subtracting, first write the numbers with the ones in the ones column, the tens in the tens column, and so on.

Example: 176 − 31 ⟶   176
                          − 31
                          ——

Then follow these steps:

Step 1:   Subtract the ones and write the answer in the ones column.

6 − 1 = 5

176
− 31
——
  5

Step 2:   Subtract the tens and write the answer in the tens column.

7 − 3 = 4

176
− 31
——
 45

Step 3:   Subtract the hundreds. In this case, there is only one number in the hundreds column. Write the 1. The answer is 145.

176
− 31
——
145

Step 4:   To check your answer, you can add from the bottom.

145 + 31 = 176

176 ⟶ 145
− 31   + 31
——   ——
145 ⟶ 176 ✓

Here's an example in which some of the digits you are subtracting are greater than the digits you are subtracting from. You need to regroup in order to subtract in that column.

1,264 − 627 ⟶   1264
                 − 627
                 ——

Step 1:   Subtract the ones. You can't subtract 7 from 4, so you need to use a number from the tens column. Regroup 6 tens and 4 ones as 5 tens and 14 ones. Write the 5 above the tens column as a reminder and write the 14 in the ones column. Now subtract 7 from 14 and write the answer in the ones column.

  5 14
1 2 6̸ 4̸
−   6 2 7
————
        7

Step 2:   Subtract the tens. Don't forget that the number in the tens column is now 5.

5 − 2 = 3

Write the answer in the tens column.

  5 14
1 2 6̸ 4̸
−   6 2 7
————
      3 7

Step 3: Subtract the hundreds. You can't subtract 6 from 2, so you need to use a number from the thousands column. Regroup 1 thousand and 2 hundreds as 12 hundreds. Cross out the 1 thousand and write the 12 in the hundreds column. Subtract 6 from 12 and write the answer in the hundreds column. There are no thousands to subtract. The answer is 637.

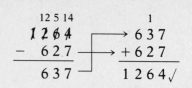

Check the answer by adding from the bottom.

637 + 627 = 1,264

To subtract larger numbers, follow the same steps. Write the number with ones lined up, tens lined up, and so on. Then subtract the ones, the tens, the hundreds, the thousands, and so on. Remember to cross out and rewrite any numbers you have to regroup.

$$
\begin{array}{r}
{\scriptstyle 8\ 13}\\
3\,4,9\,3\,9\\
-\ \ 2,8\,6\,2\\
\hline
3\,2,0\,7\,7
\end{array}
\qquad
\begin{array}{r}
{\scriptstyle 8\ 12\quad 0\ 10}\\
4\,9\,2,6\,1\,0\\
-1\,0\,9,3\,0\,9\\
\hline
3\,8\,3,3\,0\,1
\end{array}
\qquad
\begin{array}{r}
{\scriptstyle\quad\ 10}\\
{\scriptstyle 4\ 0\ 13}\\
1\,5,1\,3\,8\\
-\ \ 2,5\,5\,7\\
\hline
1\,2,5\,8\,1
\end{array}
$$

MULTIPLICATION

You can think about multiplication as repeated addition. The multiplication fact 6×5 means 6 groups of 5.

This is the same as adding the 6 groups of 5.

$$5 + 5 + 5 + 5 + 5 + 5 = 30$$
$$① \quad ② \quad ③ \quad ④ \quad ⑤ \quad ⑥$$

The answer in multiplication is called the **product.** The product of 6 and 5 is 30.

You are probably familiar with the multiplication facts—$2 \times 2 = 4$; $2 \times 3 = 6$; $2 \times 4 = 8$.... If you are unsure of any of these facts, it would be a good idea to make flash cards for yourself and memorize the facts. They are the basis of any multiplication that you do.

When you multiply, first write the numbers with the ones in the ones column, the tens in the tens column, and so on. This helps you keep track of the value of each digit as you multiply.

$$357 \times 4 \longrightarrow \begin{array}{r} 357 \\ \times\ \ 4 \\ \hline \end{array}$$

Step 1: Always multiply the **factor** on the bottom times the factor on the top. Start by multiplying the ones by the ones. $7 \times 4 = 28$ You can't write the product 28 in the ones column, so regroup as 2 tens and 8 ones. Write the 8 and write the 2 above the tens column.	 2 357 ⟵ factor $\times$ 4 ⟵ factor ‾‾‾‾‾‾ 8
Step 2: Multiply the tens by the ones. $5 \times 4 = 20$ Add the 2 that you wrote in the tens column. You can't write 22 in the tens column. Regroup.	2 2 357 $\times$ 4 ‾‾‾‾‾‾ 28
Step 3: Multiply the hundreds by the ones. $3 \times 4 = 12$ Add the 2. Write the 14. The answer is 1,428.	2 2 357 $\times$ 4 ‾‾‾‾‾‾ 1,428

Here's how to multiply by a two-digit number.

Step 1: Multiply the ones by the ones. $4 \times 6 = 24$ You can't write 24 in the ones column, so regroup as 2 tens and 4 ones. Write the 4 and write the 2 above the tens column.	2 54 $\times$ 36 ‾‾‾‾‾‾ 4
Step 2: Multiply the tens by the ones and add the 2. $5 \times 6 = 30 + 2 = 32$ Write the 32.	2 54 $\times$ 36 ‾‾‾‾‾‾ 324
Step 3: Now start multiplying by the tens. Multiply ones by tens. $4 \times 3 = 12$ The 3 in the tens place stands for 30, so it's really $4 \times 30 = 120$. Regroup. Write the 20. Write the 1 above the next column.	1 54 $\times$ 36 ‾‾‾‾‾‾ 324 20
Step 4: Multiply tens by tens and add the 1. $5 \times 3 = 15 + 1 = 16$ Write the 16.	1 54 $\times$ 36 ‾‾‾‾‾‾ 324 1,620

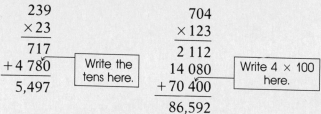

Step 5: Now add the products.

$324 \times 1620 = 1944$

$$\begin{array}{r} 54 \\ \times 36 \\ \hline 324 \\ 1\,620 \\ \hline 1{,}944 \end{array}$$

When you multiply, be careful to write each product in the right column or columns. Add the products to get the final answer.

$$\begin{array}{r} 239 \\ \times 23 \\ \hline 717 \\ +4\,780 \\ \hline 5{,}497 \end{array}$$
Write the tens here.

$$\begin{array}{r} 704 \\ \times 123 \\ \hline 2\,112 \\ 14\,080 \\ +70\,400 \\ \hline 86{,}592 \end{array}$$
Write 4×100 here.

DIVISION

You can use the multiplication facts that you know to figure out division facts.

$$3 \times 5 = 15 \longrightarrow 15 \div 5 = 3 \text{ and } 15 \div 3 = 5$$

The answer in division is called the **quotient.** Here's how to find the quotient when you divide by a one-digit number.

Step 1: Divide the hundreds by 6.

$6 \div 6 = 1$

Write the 1. Multiply 1×6.
Write the 6. Subtract.

$$\begin{array}{r} 1 \\ 6\overline{)672} \\ (1 \times 6 =)\ 6 \\ \hline 0 \end{array}$$

Step 2: Bring down the tens. Divide 7 by 6. The closest you can get to the answer is 1. Write the 1. Multiply 1×6. Write the 6. Subtract.

$$\begin{array}{r} 11 \\ 6\overline{)672} \\ 6 \\ \hline 7 \\ (1 \times 6 =)\ 6 \\ \hline 1 \end{array}$$

Step 3: Bring down the ones. Divide 12 by 6. Write the 2. Multiply 2×6. Subtract. There is no **remainder.**

$$\begin{array}{r} 112 \\ 6\overline{)672} \\ 6 \\ \hline 7 \\ 6 \\ \hline 12 \\ (2 \times 6 =)\ 12 \\ \hline 0 \end{array}$$

Here's another example.

Step 1: Try to divide the thousands. You can't divide 3 by 8. Divide the hundreds.

$31 \div 8$

The closest you can get to the answer is 3. Multiply. Subtract.

$$\begin{array}{r} 3 \\ 8\overline{)3153} \\ (3 \times 8 =)\ 24 \\ \hline 7 \end{array}$$

Step 2: Bring down the tens. Divide 75 by 8. Write 9. Multiply. Subtract.

$$\begin{array}{r} 39 \\ 8\overline{)3153} \\ 24 \\ \hline 75 \\ (9 \times 8 =)\quad 72 \\ \hline 3 \end{array}$$

Step 3: Bring down the ones. Divide 33 by 8. Write 4. Multiply. Subtract. There is 1 one left over. Write it as a remainder.

Quotient: 394 *R1*

$$\begin{array}{r} 394\ R1 \\ 8\overline{)3153} \\ 24 \\ \hline 75 \\ 72 \\ \hline 33 \\ (4 \times 8 =)\quad 32 \\ \hline 1 \end{array}$$

Here's an example in which there is a zero in the quotient.

Step 1: Divide the hundreds. Multiply. Subtract.

$$\begin{array}{r} 1 \\ 8\overline{)864} \\ (1 \times 8 =)\quad 8 \\ \hline 0 \end{array}$$

Step 2: Bring down the tens. You can't divide 6 by 8. Write a 0 in the quotient. Multiply. Subtract.

$$\begin{array}{r} 10 \\ 8\overline{)864} \\ 8 \\ \hline 6 \\ (0 \times 8 =)\quad 0 \\ \hline 6 \end{array}$$

Step 3: Bring down the ones. Divide. Multiply. Subtract. There is no remainder.

$$\begin{array}{r} 108 \\ 8\overline{)864} \\ 8 \\ \hline 6 \\ 0 \\ \hline 64 \\ (8 \times 8 =)\quad 64 \\ \hline 0 \end{array}$$

You follow the same steps when you divide by a two-digit number. Use the division facts to help you estimate.

Step 1: Try to divide the hundreds. You can't divide 3 by 62. Try the tens. You can't divide 32 by 62. Divide the ones.

$321 \div 62$

A good estimate is 5 ($5 \times 6 = 30$, so $5 \times 60 = 300$).

$$\begin{array}{r} 5 \\ 62\overline{)321} \end{array}$$

Step 2: Multiply 5×62. Subtract. Write the remainder.

$$\begin{array}{r} 5\ R11 \\ 62\overline{)321} \\ 310 \\ \hline 11 \end{array}$$

Here's another example.

Step 1: You can't divide the hundreds.
Divide the tens.

81 ÷ 24

Try 4 (4 × 2 = 8).

Multiply 4 × 24. The answer
is too large.

$$\begin{array}{r} 4 \\ 24\overline{)813} \\ (4 \times 24 =)\ \ 96 \end{array}$$

Step 2: Try a lower estimate—3. Multiply 3 × 24. Subtract.

$$\begin{array}{r} 3 \\ 24\overline{)813} \\ (3 \times 24 =)\ \ 72 \\ \hline 9 \end{array}$$

Step 3: Bring down the ones. Divide.
Multiply. Subtract. Write the
remainder.

$$\begin{array}{r} 33\ R21 \\ 24\overline{)813} \\ 72 \\ \hline 93 \\ 72 \\ \hline 21 \end{array}$$

Divide greater numbers in the same way.

Step 1: You can't divide the thousands. Divide the hundreds.

46 ÷ 34

Write the estimate. Multiply.
Subtract.

$$\begin{array}{r} 1 \\ 34\overline{)4602} \\ 34 \\ \hline 12 \end{array}$$

Step 2: Bring down the tens. Divide.

120 ÷ 34

Try 4. Too large. Try 3. Multiply. Subtract.

$$\begin{array}{r} 13 \\ 34\overline{)4602} \\ 34 \\ \hline 120 \\ 102 \\ \hline 18 \end{array}$$

Step 3: Bring down the ones.

182 ÷ 34

Try 5. Multiply. Subtract.
Write the remainder.

$$\begin{array}{r} 135\ R12 \\ 34\overline{)4602} \\ 34 \\ \hline 120 \\ 102 \\ \hline 182 \\ 170 \\ \hline 12 \end{array}$$

❏ ORDER OF OPERATIONS

Some math problems require doing more than one kind of operation.
For example, this problem requires addition, subtraction, and multiplication.

$$[4 + (10 - 7)] \times 3$$

To solve problems like this, you need to know some basic principles.

1. Do multiplication and division first. Then do addition and subtraction. Work from the left to the right.

Divide.

$2 + \boxed{16 \div 4} =$

$2 + \quad 4 \quad = 6$

Then add.

Multiply.

$3 + \boxed{8 \times 4} =$

$3 + \quad 32 \quad = 35$

Then add.

$30 - \boxed{8 \times 2} + 1 =$

$\boxed{30 - \quad 16} \quad + 1 =$

$14 \quad + 1 = 15$

$\boxed{2 \times 6} + \boxed{24 \div 8} =$

$12 \quad + \quad 3 \quad = 15$

2. When you see symbols such as [ ] or ( ), do the operations inside the symbols first.

Examples: $(3 + 8) \times 2 =$

$11 \quad \times 2 = 22$

$25 \div (3 + 2) =$

$25 \div \quad 5 \quad = 5$

3. Do the operations inside the innermost symbols first.

$[4 + (10 - 7)] \times 3 =$

$[4 + \quad 3] \quad \times 3 =$

$7 \quad \times 3 = 21$

❑ ROUNDING

Often it isn't necessary to work with exact numbers. Instead, you can make estimates using round numbers.

❑ For example, if 48 people will attend a meeting, round the number up to 50 and set up 50 chairs.

❑ If you are going to buy 11 boxes of paper costing $19.99 each, use the round number $20 for the cost and multiply by 10 to get an estimate of the total cost.

To round these numbers to tens, look at the ones position. Increase the tens digit if the ones digit is 5 or greater.

Ones digit is 5 or more.

$38 \longrightarrow 40$

Increase the tens.

$15 \longrightarrow 20 \qquad 17 \longrightarrow 20 \qquad 48 \longrightarrow 50 \qquad 76 \longrightarrow 80$

Leave the number the same if the digit is less than 5.

Ones digit is less than 5.
↓
34 ⟶ 30
↑
Stays the same.

14 ⟶ 10   22 ⟶ 20   41 ⟶ 40   75 ⟶ 80

To round to hundreds, look at the tens digit. Is it 5 or greater or less than 5?

Tens digit is 5 or more.
↓
161 ⟶ 200
↑
Round up.

Tens digit is less than 5.
↓
329 ⟶ 300
↑
Stays the same.

When you round, look to the digit to the right of the place you are rounding to.

Round to thousands:
4,⬛89 ⟶ 4000
↑
Stays the same.

Round to ten thousands:
5⬛,012 ⟶ 60,000
↑
Round up.

Round to hundred thousands:
4⬛9,228 ⟶ 400,000

Round to millions:
3,⬛21,008 ⟶ 4,000,000

❏ FRACTIONS

FRACTIONS AND MIXED NUMBERS

Fractions are parts of a whole. This pizza is cut into 8 equal pieces. Each piece is ⅛ (one-eighth) of the whole. If there are five pieces left, then ⅝ (five-eighths) of the pizza is left.

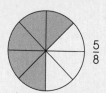

The top number of a fraction is called the **numerator.** The bottom number of a fraction is called the **denominator.** Each number has a meaning:

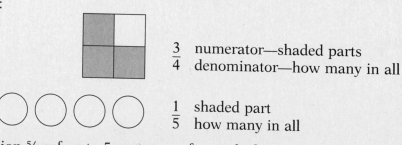

$\frac{3}{4}$ numerator—shaded parts
   denominator—how many in all

$\frac{1}{5}$ shaded part
   how many in all

The fraction ⅝ refers to 5 parts out of a total of 8 parts: 5 pizza slices out of 8, for example.

A **mixed number** includes a whole number and a fraction. For example, 1¾ is a mixed number.

You can write this mixed number as a fraction. There are ⁴⁄₄ in the square on the left. There are ³⁄₄ in the square on the right.

$$1\frac{3}{4} = \frac{7}{4} \quad \begin{array}{l} \text{shaded parts} \\ \text{how many in each whole} \end{array}$$

To change a mixed number to a fraction, follow these steps:

$$2\frac{1}{2} = ?$$

Step 1: Write the denominator. Multiply the whole number times the denominator.

$2 \times 2 = 4$

$$2 \times \overset{4}{\underset{}{}} \frac{1}{2} = \frac{}{2}$$

Step 2: Add the numerator.

$4 + 1 = 5$

$$2 \overset{4+\downarrow}{} \frac{1}{2} = \frac{5}{2}$$

To change a fraction to a mixed number, follow these steps:

Step 1: Divide the numerator by the denominator. Write the quotient as the whole number part of the mixed number.

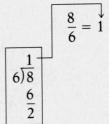

$$\frac{8}{6} = 1$$

$$\begin{array}{r} 1 \\ 6\overline{)8} \\ \underline{6} \\ 2 \end{array}$$

Step 2: Write the remainder as the numerator of the fraction part of the mixed number.

$$\frac{8}{6} = 1\frac{2}{6}$$

$$\begin{array}{r} 1 \\ 6\overline{)8} \\ \underline{6} \\ 2 \end{array}$$

EQUIVALENT FRACTIONS

There are many ways to name the same fraction, for example, ½, ⁴⁄₈, ³⁄₆ all name the same fraction. They are **equivalent fractions.**

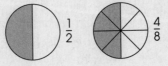

To find equivalent fractions, multiply or divide both the numerator and denominator by the same number.

$$\frac{1\ (\times\ 2)}{2\ (\times\ 2)} = \frac{2}{4}$$ The fractions ½ and ²⁄₄ are equivalent.

$$\frac{6\ (\div\ 3)}{9\ (\div\ 3)} = \frac{2}{3}$$ The fractions ⁶⁄₉ and ²⁄₃ are equivalent.

LOWEST TERMS

Fractions are **reduced to lowest terms** when you can't divide the numerator and denominator by any number other than 1 (which doesn't change the number).

$$\frac{5\ (\div\ 5)}{10\ (\div\ 5)} = \frac{1}{2}$$ This fraction is reduced to lowest terms. You can't divide both 1 and 2 by any number (except 1).

$$\frac{4\ (\div\ 2)}{8\ (\div\ 2)} = \frac{2}{4}$$ This fraction is not reduced to lowest terms. You can still divide 2 and 4 by 2. The lowest term is ½.

ADDING FRACTIONS

To add fractions, follow these steps:

Step 1: Add the numerators.

$$\frac{3}{8} + \frac{1}{8} \rightarrow = \frac{4}{\_}$$

Step 2: Write the denominator. It doesn't change.

$$\frac{3}{8} + \frac{1}{8} = \frac{4}{8}$$

Step 3: Reduce the fraction to lowest terms.

$$\frac{3}{8} + \frac{1}{8} = \frac{4\ (\div\ 4)}{8\ (\div\ 4)} = \frac{1}{2}$$

To add fractions with unlike denominators:

Step 1: Find a **common denominator** for the fractions. (This is a number that you can divide by both denominators.) One way to find a common denominator is to multiply the denominators.

$3 \times 4 = 12$

You can divide 12 by both 3 and 4.

$$\frac{2}{3} = \frac{}{12}$$

$$+\frac{3}{4} = \frac{}{12}$$

Step 2: Write equivalent fractions. Multiply the numerator by the same number that you used to find a common denominator.

$$\frac{2\ (\times\ 4)}{3\ (\times\ 4)}\ \frac{8}{12}$$

$$+\frac{3\ (\times\ 3)}{4\ (\times\ 3)}\ \frac{9}{12}$$

Step 3: Add. Write the answer in lowest terms.

$$\frac{2}{3} = \frac{8}{12}$$

$$+\frac{3}{4} = \frac{9}{12}$$

$$\frac{17}{12} = 1\frac{5}{12}$$

To add mixed numbers:

Step 1: Write equivalent fractions with common denominators.

$$2\frac{5}{8} \rightarrow 2\frac{15}{24}$$

$$+7\frac{5}{6} \rightarrow +7\frac{20}{24}$$

Step 2: Add the fractions. If the fraction can be changed to a mixed number, change it. Write the whole number part above the whole numbers.

$$2\frac{5}{8} \rightarrow 2\overset{1}{\frac{15}{24}}$$

$$+7\frac{5}{6} \rightarrow +7\frac{20}{24}$$

$$\frac{35}{24}$$

$$\frac{11}{24}$$

$$24\overline{)35} \quad \overset{1}{}$$
$$\underline{24}$$
$$11$$

Step 3: Add the whole numbers. Write the answer in lowest terms.

$$2\frac{5}{8} \rightarrow 2\overset{1}{\frac{15}{24}}$$

$$+7\frac{5}{6} \rightarrow +7\frac{20}{24}$$

$$10\frac{11}{24}$$

SUBTRACTING FRACTIONS

To subtract fractions, follow these steps:

Step 1: Subtract the numerators.

$$\frac{9}{16} - \frac{5}{16} \rightarrow \frac{4}{}$$

Step 2: Write the denominator. It doesn't change.

$$\frac{9}{16} - \frac{5}{16} = \frac{4}{16}$$

Step 3: Reduce the fraction to lowest terms.

$$\frac{9}{16} - \frac{5}{16} = \frac{4\,(\div 4)}{16\,(\div 4)} = \frac{1}{4}$$

To subtract fractions with unlike denominators:

Step 1: Write equivalent fractions with a common denominator.

$$\frac{3}{4} = \frac{12}{16}$$

$$-\frac{1}{8} = \frac{2}{16}$$

Step 2: Subtract the numerators. Write the denominator.

$$\frac{3}{4} = \frac{12}{16}$$
$$-\frac{1}{8} = \frac{2}{16}$$
$$\overline{\frac{10}{16}}$$

Step 3: Write the answer in lowest terms.

$$\frac{3}{4} = \frac{12}{16}$$
$$-\frac{1}{8} = \frac{2}{16}$$
$$\frac{10\,(\div 2)}{16\,(\div 2)} = \frac{5}{8}$$

To subtract mixed numbers:

Step 1: Write equivalent fractions with common denominators.

$$3\frac{1}{3} \rightarrow 3\frac{2}{6}$$
$$-1\frac{5}{6} \rightarrow 1\frac{5}{6}$$

Think of 3 as $2\frac{6}{6}$.

Step 2: Subtract the fractions. You can't subtract the numerator 5 from 2. Regroup 3²⁄₆ into 2 + ⁶⁄₆ + ²⁄₆ then into 2⁸⁄₆. Subtract the numerators. Write the denominator.

$$3\frac{1}{3} \rightarrow 3\frac{2}{6} \rightarrow 2\frac{6}{6} + \frac{2}{6} \rightarrow 2\frac{8}{6}$$
$$-1\frac{5}{6} \rightarrow 1\frac{5}{6} \longrightarrow 1\frac{5}{6}$$
$$\frac{3}{6}$$

Step 3: Subtract the whole numbers. Write the answer in lowest terms.

$$3\frac{1}{3} \rightarrow 2\frac{8}{6}$$
$$-1\frac{5}{6} \rightarrow 1\frac{5}{6}$$
$$\overline{1\frac{3}{6} = 1\frac{1}{2}}$$

MULTIPLYING FRACTIONS

To multiply fractions, follow these steps.

Step 1: Write the fractions horizontally, with the numerators lined up and the denominators lined up.

$$\frac{2}{5} \times \frac{5}{8} =$$

Step 2: Multiply the numerators.
$2 \times 5 = 10$
Multiply the denominators.
$5 \times 8 = 40$

$$\frac{2}{5} \times \frac{5}{8} \Rightarrow \frac{10}{40}$$

Step 3: Write the fraction in lowest terms.

$$\frac{2}{5} \times \frac{5}{8} = \frac{10\,(\div\,10)}{40\,(\div\,10)} \frac{1}{4}$$

Any whole number can be written as a fraction with 1 in the denominator.

$$3 = \frac{3}{1} \qquad 2 = \frac{2}{1} \qquad 6 = \frac{6}{1} \qquad 15 = \frac{15}{1} \qquad 7 = \frac{7}{1}$$

To multiply a whole number times a fraction:

Step 1: Write the whole number as a fraction.

$$6 \times \frac{2}{5} = \frac{6}{1} \times \frac{2}{5}$$

$$6 = \frac{6}{1}$$

Step 2: Multiply as usual.

$$6 \times \frac{2}{5} = \frac{6}{1} \times \frac{2}{5} \rightrightarrows \frac{12}{5}$$

Step 3: Change the fraction to a mixed number.

$$\frac{6}{1} \times \frac{2}{5} = \frac{12}{5} = 2\frac{2}{5}$$

$$\begin{array}{r} 2 \\ 5\overline{)12} \\ 10 \\ \hline 2 \end{array}$$

To multiply mixed numbers:

Step 1: Write the mixed numbers as fractions.

$$1\frac{1}{4} \times 2\frac{1}{2} =$$
$$\downarrow \qquad \downarrow$$
$$\frac{5}{4} \times \frac{5}{2}$$

Step 2: Multiply.

$$\frac{5}{4} \times \frac{5}{2} = \frac{25}{8}$$

Step 3: Write a mixed number for the answer.

$$\frac{5}{4} \times \frac{5}{2} = \frac{25}{8} = 3\frac{1}{8}$$

$$\begin{array}{r} 3 \\ 8\overline{)25} \\ 24 \\ \hline 1 \end{array}$$

DIVIDING FRACTIONS

When you divide fractions, turn the problem into a multiplication problem with fractions. Then multiply as usual.

Step 1: Invert the number you are dividing by. Write ⅓ as ³/₁.

$$\frac{5}{9} \div \frac{1}{3} =$$
$$\downarrow$$
$$\frac{3}{1}$$

Step 2: Multiply as usual.

$$\frac{5}{9} \div \frac{1}{3} =$$
$$\frac{5}{9} \times \frac{3}{1} = \frac{15}{9}$$

Step 3: Write the answer in lowest terms.

$$\frac{5}{9} \times \frac{3}{1} = \frac{15}{9} = 1\frac{6}{9} = 1\frac{2}{3}$$

lowest terms↓

To divide by a whole number or mixed number:

Step 1: If you have a whole number or a mixed number, write it as a fraction.

$2\frac{1}{4} = \frac{9}{4}$

$$\frac{7}{8} \div 2\frac{1}{4}$$
↓
$$\frac{7}{8} \div \frac{9}{4}$$

Step 2: Invert the number you are dividing by.

$\frac{9}{4} \longrightarrow \frac{4}{9}$

Make the numerator into the denominator.
Make the denominator into the numerator.

$$\frac{7}{8} \div \frac{9}{4}$$
↓
$$\frac{4}{9}$$

Step 3: Multiply. Write the answer in lowest terms.

$$\frac{7}{8} \times \frac{4}{9} = \frac{28\,(\div 4)}{72\,(\div 4)} = \frac{7}{18}$$

❑ DECIMALS

PLACE VALUE IN DECIMALS

Fractions where the denominator is 10 or a multiple of 10—100, 1,000, 10,000 . . . —can be written as **decimals.**

Ordinary Fraction	Decimal
$\frac{6}{10}$	0.6

Mixed Number	Decimal
$2\frac{33}{100}$	2.33

Examples:

$$\frac{3}{10} = 0.3 \qquad \frac{4}{100} = 0.04 \qquad \frac{8}{1000} = 0.008$$

This chart shows place value of the decimal numbers. The decimal point separates the whole numbers from the decimals.

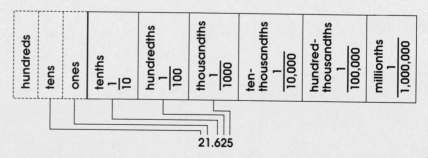

21.625

This number means:

2 tens + 1 ones + 6 tenths + 2 hundredths + 5 thousandths

To read the number, read the value of each section, using the decimal point as the divider.

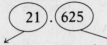

Read: twenty-one and six hundred twenty-five thousandths

To compare decimals, write the numbers with the same number of places. (You can always add zeros at the end of a decimal without changing its value.)

Compare: 0.4   and   0.01
            ↓   ↓
           40   01
       greater

Compare: 0.035   and   0.0912
             ↓   ↓
          0350   0912
                             greater

Compare: 1.06   and   1.007
             ↓   ↓
         1060   1007
       greater

ADDING DECIMALS

Adding decimals is like adding whole numbers. The only trick is to line up the decimal points.

$$6.18 + 3.24 =$$

Step 1: Write the numbers with the decimal points lined up.

       ↓
   6.18
+ 3.24

Step 2: Add. Start with hundredths.

8 + 4 = 12

You can't write 12 in the hundredths place. Regroup. Write the 2. Write the 1 over the tenths column.

       1
   6.18
+ 3.24
      2

Step 3: Add the tenths.

1 + 1 + 2 = 4

Add the ones.

6 + 3 = 9

Write the decimal point in the answer.

      1
   6 . 18
+ 3 . 24
  9↓42
  9 . 42

In adding decimals, be sure to line up the decimal points. Look at how the decimal points are lined up in each of these examples.

$$\begin{array}{r} 2.01 \\ +13.9 \\ \hline 15.91 \end{array} \qquad \begin{array}{r} 697.32 \\ +\ \ 1.048 \\ \hline 698.368 \end{array} \qquad \begin{array}{r} 7.003 \\ +0.12 \\ \hline 7.123 \end{array} \qquad \begin{array}{r} 0.02 \\ +11.796 \\ \hline 11.816 \end{array}$$

SUBTRACTING DECIMALS

In subtracting decimals, be sure to line up the decimal points.

$$7.141 - 1.028 =$$

Step 1:   Write the numbers with the decimal points lined up.

$$\begin{array}{r} 7.141 \\ -1.028 \end{array}$$

Step 2:   Subtract. Start with thousandths. You can't subtract 8 from 1. Regroup. Then subtract.

$$11 - 8 = 3$$

$$\begin{array}{r} {\scriptstyle 3\ 11} \\ 7.14\!\!\!/1 \\ -1.028 \\ \hline 3 \end{array}$$

Step 3:   Subtract the hundredths.

$$3 - 2 = 1$$

Subtract the tenths.

$$1 - 0 = 1$$

Subtract the ones.

$$7 - 1 = 6$$

Write the decimal point in the answer.

$$\begin{array}{r} {\scriptstyle 3\ 11} \\ 7.14\!\!\!/1 \\ -1.028 \\ \hline 6\ \ 113 \\ 6.113 \end{array}$$

In subtracting decimals, be sure to line up the decimal points. Look at how the decimal points are lined up in each of these examples.

$$\begin{array}{r} 17.09 \\ -\ \ 3.1 \\ \hline 13.99 \end{array} \qquad \begin{array}{r} 241.0083 \\ -\ \ \ \ 0.32 \\ \hline 240.6883 \end{array} \qquad \begin{array}{r} 9.900 \\ -3.006 \\ \hline 6.894 \end{array}$$

MULTIPLYING DECIMALS

When you multiply decimals, first multiply just as with whole numbers. Then place the decimal point.

Add the number of decimal places in each of the factors. The number of decimal places in the answer should be equal to the total number of decimal places.

Example:

$$\begin{array}{r} {\downarrow\downarrow\downarrow} \\ 2.146 \\ \downarrow \\ \times\ \ \ 1.2 \\ \hline 2.5752 \\ {\uparrow\uparrow\uparrow\uparrow} \end{array}$$

3 decimal places

$+$

1 decimal places

4 decimal places

Here are the steps to follow.

Step 1: Multiply from the right.

7 × 4 = 28

Regroup. Write the 2 in the next column.

Multiply the 1 times the 4 and add the 2 that you regrouped.

Multiply 3 × 4.

3 × 4 = 12

$$\begin{array}{r} 2 \\ 3.17 \\ \times\,0.24 \\ \hline 1268 \end{array}$$

Step 2: Multiply by the 2 tenths. Write the answer. Add.

$$\begin{array}{r} 3.17 \\ \times\,0.24 \\ \hline 1268 \\ 6340 \\ \hline 7608 \end{array}$$

Step 3: Add the decimal places in each factor.

3.17 ⟶ 2 decimal places
0.24 ⟶ 2 decimal places
4 in all

Place the decimal point with 4 decimal places in the answer.

$$\begin{array}{r} 3.17 \\ \times\,0.24 \\ \hline 1268 \\ 6340 \\ \hline 0.7608 \end{array}$$

Here's an example in which you have to write a zero after the decimal point in order to have enough decimal places in the answer.

Step 1: Multiply as usual. Add the products.

$$\begin{array}{r} 2.7 \\ \times\,0.03 \\ \hline 81 \\ 00 \\ \hline 81 \end{array}$$

Step 2: Count the total number of decimal places. There are 3 in all. Write a zero to show 3 decimal places. Write the decimal point.

$$\begin{array}{r} 2.7 \\ \times\,0.03 \\ \hline 81 \\ 00 \\ \hline .081 \end{array}$$
↓
0.081   (Write a zero in the ones place.)

DIVIDING DECIMALS

Placing the decimal point is the only thing that makes dividing decimals different from dividing whole numbers.

Follow these steps:

Step 1: Write the decimal point directly above the decimal point in the number you are dividing.

$$23\overline{)71.76}\quad\overset{.}{}$$

Step 2: Divide as usual.

$$\begin{array}{r} 3.12 \\ 23\overline{)71.76} \\ 69 \\ \hline 2\ 7 \\ 2\ 3 \\ \hline 46 \\ 46 \\ \hline 0 \end{array}$$

Example: $0.072 \div 9$

Step 1: Place the decimal point. Write zeros above the zeros.

$$\begin{array}{r} 0.0 \\ 9\overline{)0.072} \end{array}$$

Step 2: You can't divide 7 by 9. Write a zero in the answer. Then divide the thousandths.

$$\begin{array}{r} 0.008 \\ 9\overline{)0.072} \\ 72 \\ \hline 0 \end{array}$$

To divide *by* a decimal, follow these steps:

$$1.176 \div 0.21 =$$

Step 1: Move the decimal point in the number you are dividing by. Move the decimal point in the number you are dividing the same number of places.

$$0.21\overline{)1.176}$$

Step 2: Write the decimal point.

$$21\overline{)117.6}$$

Step 3: Divide as usual.

$$\begin{array}{r} 5.6 \\ 21\overline{)117.6} \\ 105 \\ \hline 12\ 6 \\ 12\ 6 \\ \hline 0 \end{array}$$

ROUNDING DECIMALS

To round a decimal number, look at the digit in the place to the right of the place you want to round to.

Look at hundredths.

Round to tenths: ↓
          3.16

Look at thousandths.

Round to hundredths: ↓
             3.245

If the digit is 5 or greater, round up.
If the digit is less than 5, leave the same.

	look		look	
	↓	up	↓	same
Round to tenths:	3.16 ⟶ 3.2		7.12 ⟶ 7.1	
Round to hundredths:	3.242 ⟶ 3.24		0.008 ⟶ 0.01	
	↑	same	↑	up
	look		look	

ROUNDING DECIMAL ANSWERS
When you divide decimals, don't show a remainder.

Example: Divide 5.99 by 43. Show your answer in hundredths.

Step 1: Write the decimal point in the answer. Divide as usual.

$$\begin{array}{r} 0.13 \\ 43\overline{)5.99} \\ 43 \\ \hline 1\,69 \\ 1\,29 \\ \hline 40 \end{array}$$

Step 2: The answer should be rounded to hundredths. Divide to one place beyond hundredths. There is no digit in the thousandths place. Add a zero. Divide thousandths.

$$\begin{array}{r} 0.139 \\ 43\overline{)5.990} \\ 43 \\ \hline 1\,69 \\ 1\,29 \\ \hline 400 \\ 387 \\ \hline 13 \end{array}$$

Step 3: Round the answer to hundredths.

look
↓
$$\begin{array}{r} 0.139 \quad 0.14 \\ 43\overline{)5.990} \end{array}$$

When you divide whole numbers, you can show the remainder as a decimal.

Example: Divide 200 by 6. Carry the division to tenths.

Step 1: Divide as usual. When you get a remainder, write a decimal point. Add zeros in the tenths and hundredths places. Continue to divide.

$$\begin{array}{r} 33.33 \\ 6\overline{)200.00} \\ 18 \\ \hline 20 \\ 18 \\ \hline 2\,0 \\ 1\,8 \\ \hline 20 \\ 18 \\ \hline 2 \end{array}$$

Step 2: Round your answer to tenths.

look
↙
$$\begin{array}{r} 33.33 \quad 33.3 \\ 6\overline{)200.00} \end{array}$$

❏ PERCENTS, RATIOS, AND PROPORTIONS

RATIOS

A comparison of two quantities is called a **ratio.** A fraction is a ratio. In this diagram, there are 4 stars and 8 dots. The ratio of stars to dots is 4 to 8.

The ratio can be written:

$$4 \text{ to } 8 \qquad 4:8 \qquad \text{or} \qquad \frac{4}{8}$$

The two numbers in a ratio are the **terms** of the ratio.

As with a fraction, a ratio can be written in lowest terms. The fraction ⁴⁄₈ can be written as ½. Write the ratio as:

$$1 \text{ to } 2 \qquad 1:2 \qquad \text{or} \qquad \frac{1}{2}$$

Rectangle A has sides of 1 foot and 3 feet. The ratio of the short side to the longer side is 1:3. Rectangle B has sides of 2 feet and 6 feet. The ratio of the short side to the long side is ²⁄₆ or ⅓. The ratios are **equal**.

6 ft

3 ft

1 ft

2 ft

RECTANGLE A

RECTANGLE B

To find an equal ratio, multiply each term by the same number. These ratios are equal:

$$\frac{1 \ (\times 2)}{2 \ (\times 2)} = \frac{2}{4} \qquad \frac{1 \ (\times 5)}{2 \ (\times 5)} = \frac{5}{10} \qquad \frac{1 \ (\times 8)}{2 \ (\times 8)} = \frac{8}{16}$$

If the sides of rectangle A are kept in the same ratio, and the short side is enlarged to 6 feet, how long will the longer side be? Set up equal ratios and find the missing number.

$$\frac{1}{3} = \frac{6}{?} \qquad \frac{1 \ (\times 6)}{3 \ (\times 6)} = \frac{6}{18}$$

(Think: What number was 1 multiplied by?)

(Multiply 3 by that number.)

The longer side will be 18 feet long.

PROPORTIONS

A **proportion** says that two ratios are equal.

Examples of proportions: $\dfrac{2}{3} = \dfrac{4}{6} \qquad \dfrac{2}{10} = \dfrac{1}{5} \qquad \dfrac{1}{2} = \dfrac{4}{8}$

In a proportion, the cross products are equal.

$6 \times 1 = 6$
$3 \times 2 = 6$

$6 = 6$ The ratios are a proportion because the cross products are equal.

To find cross products, multiply the upper term of one ratio times the lower term of the other. Multiply the lower term times the upper term.

Use cross products to solve problems involving proportions.

Problem: It takes 2 eggs to make pancakes for 3 people. How many eggs will it take to make pancakes for 12 people?

Step 1: Write a proportion. Use n for the unknown number. Be sure the ratios name the items in the same order.

$$\frac{\text{egg}}{\text{people}} = \frac{\text{egg}}{\text{people}}$$

$$\frac{2}{3} = \frac{n}{12}$$

Step 2: Multiply to get the cross products. Set them equal to each other.

$$\frac{2}{3} \times \frac{n}{12}$$

$$3n = 24$$

Step 3: To solve the equation, divide each side by the same number. The missing number is 8 eggs.

$$3n = 24$$
$$\div 3 \quad \div 3$$
$$\downarrow \quad \downarrow$$
$$n = 8$$

You can solve the problem with equal ratios instead. Follow these steps:

Step 1: Write the proportion.

$$\frac{2}{3} = \frac{n}{12}$$

Step 2: Decide what number the lower term was multiplied by. Multiply the upper term by the same number.

$$\frac{2\,(\times 4)}{3\,(\times 4)} = \frac{8}{12}$$

$$n = 8$$

PERCENTS

A **percent** is a ratio where the second term is 100. Percents are written with the percent sign (%). These three ratios are equal. The third is a percent.

$$59:100 \qquad 59 \text{ to } 100 \qquad 59\% \qquad \text{(Say 59 percent.)}$$

A percent can be written as a fraction with 100 in the denominator. It can also be written as a decimal.

$$50\% = \frac{50}{100} = 0.50 \qquad \text{(The number 0.50 means 50 hundredths or } {}^{50}\!/_{100}.)$$

To change a percent to a decimal, move the decimal point two places to the left, and drop the percent sign.

$$37\% = 0.37 \qquad 115\% = 1.15 \qquad 3567\% = 35.67$$
(These are the steps: $37\% = {}^{37}\!/_{100} = 0.37$.)

You can also change any decimal to a percent.

To change a decimal to a percent, move the decimal point two places to the right and add the percent sign.

$$0.89 = 89\% \qquad 1.73 = 173\% \qquad 0.01 = 1\%$$
(These are the steps: $0.89 = {}^{89}\!/_{100} = 89\%$.)

You can write any ratio as a percent. Use cross products.

One out of ten people is left-handed. What percent is that? (Remember, a percent has 100 in the denominator.)

$$\frac{1}{10} = \frac{n}{100} \qquad \begin{array}{l} 10n = 100 \\ \div 10 \quad \div 10 \\ n = 10 \end{array} \qquad 10 \text{ percent}$$

This formula can be used to solve percentage problems.

percent $\times$ base = amount

Examples of the formula:

$25\% \times 8 = 2$ $30\% \times 100 = 30$

(25 percent of 8 is 2.)   (30 percent of 100 is 30.)

$15\% \times 60 = 9$ $75\% \times 200 = 150$

(15 percent of 60 is 9.)   (75 percent of 200 is 150.)

If you know two of the numbers in the formula, you can find the third.

FINDING THE AMOUNT

To find 30 percent of 500, follow these steps:

Step 1: Write the numbers you know and the unknown number in the percent formula. (Think 30 percent of 500 is ____.)

percent $\times$ base = amount
$30\% \times 500 = n$

Step 2: Solve the formula. Multiply. To multiply by a percent, change it to a decimal first.

$30\% \times 500 = n$
$\downarrow$
$.30 \times 500 = n$

$$\begin{array}{r} 500 \\ \times \ .30 \\ \hline 150.00 \end{array}$$

FINDING THE BASE

To solve a problem such as "12 percent of what number is 3.6?" follow these steps.

Step 1: Write the numbers you know and the unknown number in the percent formula. This time, you know the amount (3.6) and you know the percent (12).

(Think 12 percent of ____ is 3.6.)

percent $\times$ base = amount
$12\% \times n = 3.6$

Step 2: To solve the equation, divide each side by the same number—12 percent. First convert 12 percent to a decimal.

$12\% \times n = 3.6$
$\downarrow$
$.12 \times n = 3.6$

Step 3: Divide.

$$\begin{array}{r} .12 \times n = \ \ 3.6 \\ \div \ .12 \quad \div \ .12 \\ n = 30 \end{array}$$

FINDING THE PERCENT

To solve a problem such as "What percent is 16 of 64?" follow these steps:

Step 1: Write the numbers you know and the unknown number in the percent formula.

(Think ____ percent of 64 is 16.)

$$\text{percent} \times \text{base} = \text{amount}$$
$$n \times 64 = 16$$

Step 2: Divide each side by the same number. Change the decimal to a percent.

$$n \times 64 = 16$$
$$\div 64 \qquad \div 64$$
$$n = .25$$
$$25\%$$

$$\begin{array}{r} .25 \\ 64\overline{)16.00} \\ 12\ 8 \\ \hline 3\ 20 \\ 3\ 20 \end{array}$$

FINDING PERCENT OF INCREASE OR DECREASE

Many percent problems involve an increase or decrease. For example, a quantity has increased from 40 to 60. What percent of increase was that?

Follow these steps to find a percent of increase:

Step 1: First find the amount of increase.

$$60 - 40 = 20$$

The amount of increase is 20.

$$\begin{array}{r} 60 \\ -40 \\ \hline 20 \end{array}$$

Step 2: Write the numbers you know in the formula. Use the original number as the base. Divide to solve.

$$\text{percent} \times \text{base} = \text{amount}$$
$$n \times 40 = 20$$
$$\div 40 \qquad \div 40$$
$$n = .5$$

Step 3: Convert the decimal to a percent. The answer is that 40 to 60 is a 50-percent increase.

$$.50 = 50\%$$

Suppose a quantity has decreased from 60 to 40. What is the percent of decrease from 60 to 40?

Follow these steps to find a percent of decrease:

Step 1: First determine how much the quantity decreased.

$$60 - 40 = 20$$

$$\begin{array}{r} 60 \\ -40 \\ \hline 20 \end{array}$$

Step 2: Write the numbers you know in the formula. Use the original number as the base. Divide out to thousandths. Round to hundredths.

$$\text{percent} \times \text{base} = \text{amount}$$
$$n \times 60 = 20$$
$$\div 60 \qquad \div 60$$
$$n = .333$$

Step 3: Convert the decimal to a percent. The answer is that 60 to 40 is a 33-percent decrease.

$$.33 = 33\%$$

DISCOUNT AND MARKUP

Many problems with percents have to do with discounts and markups.

Example: An item is marked "25% off." The original price is $100. What is the new price?

Follow these steps:

Step 1: First find the discount. The discount is the amount that the price will be reduced. It is 25 percent of the original price.

(25 percent of $100 is ____.)

$$percent \times base = amount$$
$$25\% \times 100 = n$$
$$.25 \times 100 = n$$
$$25 = n$$

Step 2: Subtract the discount from the original price.

$$\begin{array}{r} \$100 \\ -25 \\ \hline \$75 \end{array}$$

Example: An item costs $25. The store selling it will mark up the price by 15 percent. What will the selling price be?

Follow these steps:

Step 1: The markup is the amount the price will be raised. It is 15 percent of the original price.

(15 percent of $25 is ____.)

$$percent \times base = amount$$
$$15\% \times 25 = n$$
$$.15 \times 25 = n$$
$$3.75 = n$$

Step 2: Add the markup to the original price.

$$\begin{array}{r} \$25.00 \\ +\$3.75 \\ \hline \$28.75 \end{array}$$

❏ DESCRIPTIVE STATISTICS

Descriptive statistics are ways to describe data. It's often useful to describe data in two ways. One is by **measures of central tendency** (how data group around certain numbers). Other useful descriptions are **measures of variability** (how data are spread or scattered).

❏ Example of a statement that is a measure of central tendency: "The average number of students per class is 27."

❏ Example of a statement that is a measure of variability: "The largest class has 45, and the smallest class has 16."

MEASURES OF CENTRAL TENDENCY

Mean   The sum of all the data divided by the number of data. (Sometimes called the "average.")

Median   The middle number when the data are arranged in numerical order.

Mode   The number that occurs most often.

Here is how to find the mean, the median, and the mode for this set of data.

$$84 \quad 86 \quad 84 \quad 87 \quad 92 \quad 95 \quad 67$$

To find the mean:

Step 1: Add the numbers.

$$84 + 86 + 84 + 87 + 92 + 95 + 67 = 595$$

Step 2: Count the data. There are 7 in all. Divide by the number of data.

$$\begin{array}{r} 85 \\ 7\overline{)595} \\ \underline{56} \\ 35 \end{array}$$

$$\text{mean} = 85$$

To find the median:

Step 1: Arrange the numbers in order from least to greatest.

$$67 \quad 84 \quad 84 \quad 86 \quad 87 \quad 92 \quad 95$$

Step 2: Find the middle number. The number of data is 7, so the fourth number is the middle one.

$$67 \quad 84 \quad 84 \quad 86 \quad 87 \quad 92 \quad 95$$
$$\uparrow$$
$$\text{median} = 86$$

To find the median when you have an even number of data:

Step 1: Arrange the numbers in order and look for the middle number. There are 8 numbers, so no score falls right in the middle. Pick the 2 numbers just above and below the middle.

$$9 \ 10 \ 10 \ 12 \ 14 \ 17 \ 17 \ 19$$
$$12 \ 14$$

Step 2: Find the mean (average) of these 2 numbers. Add them. Divide by 2.

$$12 + 14 = 26$$
$$26 \div 2 = 13$$
$$\text{median} = 13$$

To find the mode:

Step 1: It usually helps to arrange the data in order. Then count to see how many of each number you have.

$$67 \quad 84 \quad 84 \quad 86 \quad 87 \quad 92 \quad 95$$
$$1 \qquad \quad 1 \quad \ 1 \quad \ 1 \quad \ 1$$
$$2$$

Step 2: The number that occurs most often is the mode.

$$\text{mode} = 84$$

MEASURES OF VARIABILITY

The most common measure of variability is range.

Range The difference between the highest and the lowest numbers in a group of data.

Step 1: Again, it helps to arrange the data in order. Find the lowest and the highest number.

$$67 \quad 84 \quad 84 \quad 86 \quad 87 \quad 92 \quad 95$$

Step 2: Subtract the lowest number from the highest number. The difference is the range.

$$95 - 67 = 28$$
$$\text{range} = 28$$

❏ GRAPHS AND TABLES

Graphs and tables organize and present information. Things that can be counted and that can be divided in some way can be shown in a graph or table.

A table contains numbers arranged in rows and columns. The rows and columns have labels. This table shows how much steel a factory produced over a 5-month period.

Example of a Completed Table:

TABLE — FACTORY STEEL PRODUCTION

MONTH	TONS OF STEEL
JAN	23
FEB	40
MAR	26
APR	30
MAY	40

A graph shows information visually. This graph shows the same information.

Example of a Completed Graph:

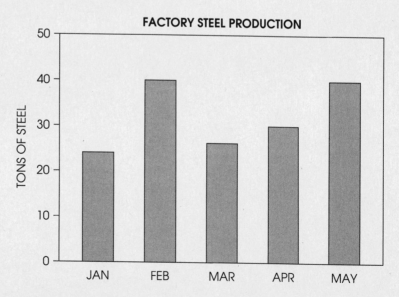

The graph lets you make comparisons easily. The tallest bar shows the greatest production. The smallest bar shows the least production. The graph also lets you see trends. There is a big drop in month 3 and then a return in month 5 to the highest level of production.

The table lets you see the numbers clearly. Comparisons and trends are not as clear. Often you make a table before you make a graph to help organize the numbers first.

Different kinds of graphs are shown below.

Examples of Different Kinds of Graphs:

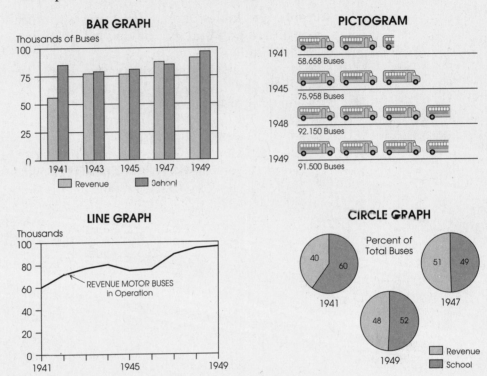

BAR GRAPHS

Bar graphs are used to show comparison. A bar graph has a **vertical axis** and a **horizontal axis.** It has bars showing data.

Example of a Completed Bar Graph:

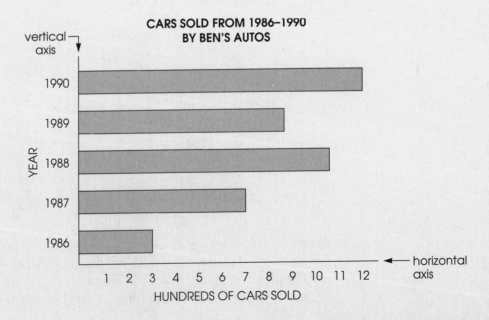

This graph shows the number of cars sold by a dealer from 1986 to 1990.

To read any bar graph, follow these steps:

❏ Read the title of the graph to see what the graph is about. This graph shows car sales for a dealer.

❏ Read the label on each axis. The vertical axis shows years from 1986 through 1990. The horizontal axis shows cars sold. Notice that each number of the horizontal axis stands for that many hundreds of cars.

❏ To read any data on the graph, find the label and the bar that goes with it. Then trace a line from the end of the bar to the other axis.

❏ To see how many cars the dealer sold in any year, find the label for the year. The bar stands for the cars sold that year. For 1986, the end of the bar is at 3, which stands for 300 cars. For 1989, the end of the bar is three-fourths of the way from 8 to 9 or about 8.75. It stands for 875 cars.

❏ The longest bar is the greatest number. The shortest bar is the least number.

The year with fewest sales was 1986.
The year with most sales was 1990.

LINE GRAPHS

Line graphs are used to show trends over time. A line graph also has two axes. It shows data with a line.

Example of a Completed Line Graph:

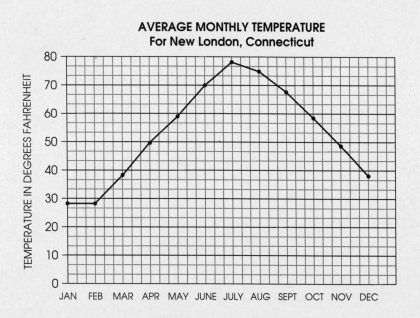

This graph shows temperatures.

To read any line graph, follow these steps:

❏ Read the title of the graph to see what the graph is about. This graph shows the average monthly temperature for New London, Connecticut.

❏ Read the label on each axis. The vertical axis shows temperature in degrees from 0 to 80. Notice that only every tenth degree is labeled. The horizontal axis shows months.

❏ To read the data, find the correct dot. Trace from the dot to the axis to read the data. To find the temperature in May, find the label "May." Locate the dot above May. Trace it to the vertical axis. It is at just below 60, or about 58° F. The dot for April is at 50° F.

❏ The line shows trends up and down. The labels on the graph tell when the trend started up or down. The average monthly temperature rises from February through July and then falls to the end of the year.

CIRCLE GRAPHS

Circle graphs are used to compare parts of a whole. A circle graph is a circle, with data shown as parts (wedges) of the circle.

Example of a Completed Circle Graph:

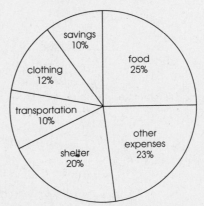

HOW A TYPICAL AMERICAN FAMILY SPENDS ITS INCOME

This graph shows how income is spent.

To read any circle graph, follow these steps:

❏ Read the title of the graph to see what the graph is about. This graph shows how a typical family spends its income.

❏ Read the label on each section of the circle. It identifies the section and gives a percentage. The graph shows the percentage spent on clothing, savings, food, and so on.

❏ A circle graph shows how a total quantity is divided. The sections of the graph always total 100 percent. To read the data, find the correct section. Read the percentage in the section. To find the percent of total income that a family spends on savings, find the savings section. The percentage shown is 10 percent.

❏ The circle graph shows comparisons between sizes of items. The sections for transportation and savings are the same. Each is 10 percent. The largest section is food, at 25 percent. Even if the graph labels don't include percentages, you can compare the sections. You can estimate how large each is.

❏ GEOMETRY

There are many different kinds of geometric figures. Here are some common ones:

parallelogram A figure with four sides; the opposite sides are equal and parallel to each other.

rectangle A special kind of parallelogram, with four right angles. Opposite sides are equal and parallel.

square A special kind of rectangle, with four equal sides and four right angles.

triangle A figure with three sides.

right triangle A triangle with one right angle.

PERIMETER

The **perimeter** is the total distance around a figure. It is the sum of the sides.

To find the perimeter of a triangle, add the lengths of the sides.

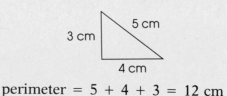

perimeter = 5 + 4 + 3 = 12 cm

To find the perimeter of a parallelogram or a rectangle, add the lengths of the sides.

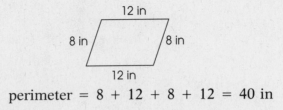

$$\text{perimeter} = 8 + 12 + 8 + 12 = 40 \text{ in}$$

You can find the perimeter of a parallelogram or rectangle if you know the length of one shorter side and one longer side.

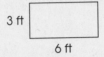

The opposite sides of a parallelogram or rectangle are equal. So you know the other two sides are also 6 feet and 3 feet.

$$\text{perimeter} = 6 + 3 + 6 + 3 = 18 \text{ ft}$$

Since the sides of a square are equal, you can find the perimeter by multiplying 4 times the length of any side.

15 ft

$$\text{perimeter} = 4 \times 15 = 60 \text{ ft}$$

AREA

Area is the number of square units needed to cover a figure. A square unit is a square inch (sq in), square centimeter (sq cm), square yard (sq yd), square mile (sq mi), and so on.

A square centimeter is 1 centimeter on a side. A square inch is one inch on a side. A square yard is 1 yard on a side.

SQUARE CENTIMETER

This rectangle is covered by 8 square feet. Its area is 8 square feet.

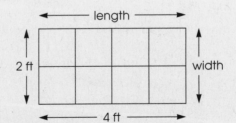

To find the area of a rectangle, use this formula: $A = l \times w$

A stands for *area*; l stands for *length*; w stands for *width*.

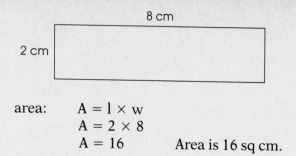

area: $A = l \times w$
$A = 2 \times 8$
$A = 16$ Area is 16 sq cm.

To find the area of a square, again multiply the length times the width. Since the sides of a square are equal, you multiply the length of one side times itself.

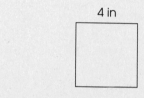

area: $A = s \times s$
$A = 4 \times 4$
$A = 16$ Area is 16 sq in.

This diagram shows a rectangle cut in half. Each half is a triangle. The area of each half of the rectangle is equal. Looking at this diagram helps you to understand how to find the area of a triangle.

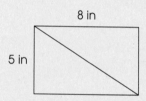

The area of the rectangle is: $A = l \times w$
$A = 5 \times 8 = 40$ sq in

The area of each triangle is half of this, or 20 square inches.

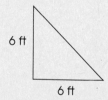

One way to write the formula for the area of a triangle is:

$$A = \frac{1}{2} \times (l \times w) \leftarrow \text{half the area of a rectangle}$$

$$A = \frac{1}{2} \times 6 \times 6 = \frac{1}{2} \times 36 = 18 \text{ sq ft}$$

This formula works because the triangle is a right triangle, and it is half of a rectangle. What about other triangles? The usual formula uses the base of the triangle and its height.

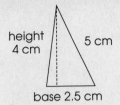

height
4 cm

5 cm

base 2.5 cm

The shortest side of a triangle is the base. The shortest distance from the base to the highest point of the triangle is the height.

$$A = \frac{1}{2}(b \times h)$$

$$A = \frac{1}{2}(2.5 \times 4)$$

$$A = \frac{1}{2}(10) = 5 \qquad \text{Area is 5 sq cm.}$$

The formula for the area of a parallelogram is:

$$A = b \times h$$

height
10 in

20 in

$$A = b \times h$$
$$A = 20 \times 10 = 200 \qquad \text{Area is 200 sq in.}$$

CIRCLES

A circle is a closed curve with all points equally distant from the center. The distance from the center of the circle to the rim of the circle is the **radius** of the circle.

radius

The distance around a circle is the **circumference.** The formula for circumference uses the Greek symbol π, which is spelled *pi*. It is pronounced "pie." Pi always has the same value.

$$\pi = \text{approximately 3.14 or } \frac{22}{7}$$

The formula for circumference of a circle is:

$$C = 2 \times \pi r$$

C is the circumference; *r* is the radius.

The circumference of the circle is: $C = 2 \times \pi \times 5$
$C = 2 \times 3.14 \times 5$
$C = 31.4$ in

The formula for area of a circle is: $A = \pi \times r \times r$

The area of the circle above is: $A = \pi \times r \times r$
$A = 3.14 \times 5 \times 5$
$A = 78.5$ sq in

VOLUME

Squares, rectangles, triangles, and circles are flat figures. Figures that have three dimensions are **solid figures.** Each surface of a solid figure is a **face.**

Prism   A solid figure with parallel bases. The other faces are parallelograms.

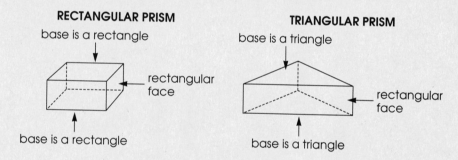

Cube   A solid figure with square bases and faces.

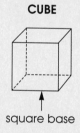

Cylinder   A solid figure with two bases that are circles.

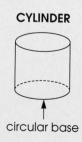

The volume of a solid figure is the number of cubic units it holds. A cubic unit is a cubic inch (cu in), cubic centimeter (cu cm), and so on.

CUBIC CENTIMETER

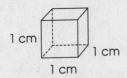

To find the volume of a prism, use this formula:

$$V = l \times w \times h$$

V is volume; *l* is length; *w* is width; *h* is height.

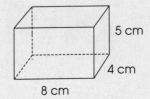

$V = l \times w \times h$
$V = 8 \times 4 \times 5 = 160$ Volume is 160 cu cm.

To find the volume of a cube, use the same formula. The length, width, and height of a cube are all the same.

$V = l \times w \times h$
$V = 6 \times 6 \times 6 = 216$ Volume is 216 cu in.

To find the volume of a cylinder, first find the area of the base and then multiply times the height. The formula is:

$$V = (\pi \times r \times r) \times h$$

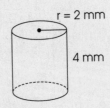

$V = (\pi \times r \times r) \times h$
$V = (3.14 \times 2 \times 2) \times 4$
$V = 12.56 \times 4 = 50.24$ Volume is 50.24 cu mm.

Mathematics Review
Skills Table

First, use the Answer Key on page 185 to correct your test. Then, fill in an X in the second column next to each answer that is wrong. Look at the page numbers in the third column next to each X. The third column tells you which pages you should study to help build your skills.

Question	Put an **X** for a wrong answer	Study these pages
1		Pages 145 – 155
2		
3		
4		
5		Pages 155 – 161
6		
7		
8		
9		
10		Pages 161 – 166
11		
12		
13		Pages 166 – 171
14		
15		
16		
17		Pages 171 – 173
18		Pages 177 – 182
19		
20		

Answer Key for Math Pretest

1. 1,462 copies
2. 4,675 pages
3. 720 hours
4. 15 pages
5. 5⅛ in.
6. 2⁷⁄₁₀ in.
7. 2½ ft
8. 13 columns
9. 1 ounce = $0.29, 3 ounces = $0.75, 7 ounces = $1.67, and 11 ounces = $2.59
   Total postage = $5.30
10. $4.74
11. $5,323.55
12. 0.10
13. 2,400 pages
14. 25%
15. 15%
16. $14.80
17. mean = $8.18
    median = $8.45
    mode = $8.45
18. 36 ft
19. 324 square ft
20. 4,275 cubic ft

PART 4

Answer Key

Job Situation 1

TASK 1

1. **To Do**: Record the check in the register, and find the new balance.

   Steps: 1. Record all the information about the current check.

   2. Subtract the new check amount from the previous balance.

   Information: Wing Construction Company; January 10; check number 7801; amount $786.50

   System of Measurement: Money

   Computation: $5,152.50 − $786.50 = $4,366

   Communicate: See completed checkbook register on page 187.

2. **To Do:** Record the check in the register, and find the new balance.

   Steps: 1. Record all the information about the current check.

   2. Subtract the new check amount from the previous balance.

   Information: Kramer's Air Conditioning; January 12; $779.37; check number 7802.

   System of Measurement: Money

   Computation: $4,366 − $779.37 = $3,586.63

   Communicate: See completed checkbook register on page 187.

3. **To Do:** Record the check in the register, and find the new balance.

   Steps: 1. Record all the information about the current check.

   2. Subtract the new check amount from the previous balance.

   Information: First Federal; January 13; check number 7803; $1,010

   System of Measurement: Money

   Computation: $3,586.63 − $1,010 = $2,576.63

   Communicate: See completed checkbook register on page 187.

TASK 2

4. **To Do:** Record the deposit in the register, and find the new balance.

   Steps: 1. Record all the information about the deposit.

   2. Add the deposit to the previous balance.

   Information: January 15; deposit of $11,179.02; from Security Envelope.

   System of Measurement: Money

   Computation: $6,076.63 + $11,179.02 = $17,255.65

   Communicate: See completed checkbook register on page 187.

5. **To Do:** Record the deposit in the register, and find the new balance. Record the information about the check, and find the new balance.

   Steps: 1. Record all the information about the deposit.

   2. Add the deposit to the previous balance.

   3. Record the information about the check.

   4. Subtract to find the new balance.

Information: January 16; deposit of $525.25; from Wilson Stationery. Check number 7804; to Jane Howard; for $1,200.

System of Measurement: Money

Computation: $17,255.65 + $525.25 = $17,780.90
$17,780.90 − $1,200 = $16,580.90

Communicate: See completed checkbook register below.

6. **To Do:** Record the information about the check and find the new balance. Record the deposit in the register and find the new balance.

    Steps: 1. Record the information about the check.

           2. Subtract to find the new balance.

           3. Record all the information about the deposit.

           4. Add the deposit to the previous balance.

Information: January 17; check for $50; to B&B Office Supplies; check number 7805. Deposit of $514.75; from Party Tyme Supplies.

System of Measurement: Money

Computation: $16,580.90 − $50 = $16,530.90
$16,530.90 + $514.75 = $17,045.65

Communicate: See completed checkbook register below.

Check No.	Date	Transaction or Payee	Payment	Deposit	Balance
7799	1/02/9X	Hardik's Office Supply	650.00		4,250.00
	1/05/9X	Payment — Olsen's		1,250.00	5,500.00
7800	1/10/9X	Northern Electric	347.50		5,152.50
7801	1/10/9X	Wing Construction Co.	786.50		4,366.00
7802	1/12/9X	Kramer's Air Cond.	779.37		3,586.63
7803	1/13/9X	First Federal	1,010.00		2,576.63
	1/14/9X	Payment — Jackson Stationers		3,500.00	6,076.63
	1/15/9X	Payment — Security Envelope		11,179.02	17,255.65
	1/16/9X	Payment — Wilson Stationery		525.25	17,780.90
7804	1/16/9X	Jane Howard	1,200.00		16,580.90
7805	1/17/9X	B&B Office Supplies	50.00		16,530.90
	1/17/9X	Payment — Party Tyme Supplies		514.75	17,045.65

TASK 3

7. **To Do:** Find the adjusted balance for the check register.

    Steps: 1. Find the total of the additions to the account.

           2. Add the additions to the balance.

           3. Find the total of the deductions to the account.

           4. Subtract the deductions from the balance.

Information: Bank fee is $8. Check register shows $1,830.50.

System of Measurement: Money

Computation:

Register balance	$1,830.50
Deductions	8.00
	$1,822.50

Communicate:

Check No.	Date	Transaction or Payee	Payment	Deposit	Balance
6328	11/30/9X	Jackson Electric	$423.00		$1,830.50
	11/30/9X	Bank Fees (November)	8.00		$1,822.50

8. **To Do:** Find the adjusted balance for the check register.

  Steps: 1. Find the total of additions to the account.

  2. Add the additions to the balance.

  3. Find the total of deductions to the account.

  4. Subtract the deductions from the balance.

Information: Bank fee is $7. Credit of $7.50. Balance is $5,235.27.

System of Measurement: Money

Computation:

Register balance	$5,235.27
Additions	7.50
	$5,242.77
Deductions	7.00
	$5,235.77

Communicate:

Check No.	Date	Transaction or Payee	Payment	Deposit	Balance
6333	12/28/9X	Federal Express	$221.00		$5,235.27
	12/28/9X	Bank Refund		7.50	5,242.77
	12/28/9X	Bank Fees	7.00		5,235.77

TASK 4

9. **To Do:** Find the adjusted balance for the bank statement. Compare it to the adjusted bank balance.

  Steps: 1. Find the total of unrecorded deposits.

  2. Add the total of deposits to the bank statement balance.

  3. Find the total of outstanding checks.

  4. Subtract the total of outstanding checks.

  5. Compare the adjusted balances.

Information:

Statement balance: $1,445.50

Unrecorded deposits shown in the check register: $800

Outstanding checks shown in the check register: $423

System of Measurement: Money

Computation:

Total unrecorded deposits:	$ 800.00
+ Statement balance:	$1,445.50
Subtotal	$2,245.50
- Total outstanding checks:	$ 423.00
Adjusted Balance	$1,822.50

Communicate:

Reconciliation Worksheet	
A. Ending Balance on Bank Statement:	$ 1,445.50
B. Deposits Made After Last Entry on Statement:	$ 800.00
C. Subtotal (Add A + B.):	$ 2,245.50
D. Total Outstanding Checks:	$ 423.00
Total (Subtract D from C.)	$ 1,822.50

(This amount should equal your checkbook balance.)
Account is reconciled.

10. **To Do:** Find the adjusted balance for the bank statement. Compare it to the adjusted bank balance.

   Steps: 1. Find the total of unrecorded deposits.

   2. Add the total of deposits to the bank statement balance.

   3. Find the total of outstanding checks.

   4. Subtract the total of outstanding checks.

   5. Compare the adjusted balances.

Information:

Statement balance: $5,758.77

Unrecorded deposits shown in the check register:	$1,198.00
Outstanding checks shown in the check register:	$1,500.00
	+221.00
	$1,721.00

System of Measurement: Money

Computation:

Total unrecorded deposits:	$1,198.00
+ Statement balance:	$5,758.77
Subtotal	$6,956.77
- Total outstanding checks:	$1,721.00
Adjusted Balance	$5,235.77

Communicate:

Reconciliation Worksheet	
A. Ending Balance on Bank Statement:	$ 5,758.77
B. Deposits Made After Last Entry on Statement:	$ 1,198.00
C. Subtotal (Add A + B.):	$ 6,956.77
D. Total Outstanding Checks:	$ 1,721.00
Total (Subtract D from C.)	$ 5,235.77

(This amount should equal your checkbook balance.)
Account is reconciled.

JOB SITUATION 2

TASK 1

1. **To Do:** Find Barbara Lopez's gross pay. Write the total hours, regular pay, and gross pay in the payroll register.

    Steps: 1. Find the total number of hours worked.

    2. Find the regular pay (gross pay).

    Information: Regular Pay Rate: $8.45 per hour.

    Hours Worked:

Monday	8 hours
Tuesday	8 hours
Wednesday	8 hours
Thursday	8 hours
Friday	8 hours

    System of Measurement: Time and money

    Computation:

    $8 + 8 + 8 + 8 + 8 = 40$ hours
    $40 \times \$8.45 = \$338$

    Regular pay is the gross pay.

    Communicate: See completed payroll register on page 193.

2. **To Do:** Find Susan Marino's gross pay. Write the total hours, regular pay, and gross pay in the payroll register.

    Steps: 1. Find the total number of hours worked.

    2. Find the regular pay (gross pay).

    Information: Regular Pay Rate: $8.40 per hour.

    Hours Worked:

Monday	8 hours
Tuesday	8 hours
Wednesday	8 hours
Thursday	7 hours
Friday	5 hours

    System of Measurement: Time and money

    Computation:

    $8 + 8 + 8 + 7 + 5 = 36$ hours
    $36 \times \$8.40 = \$302.40$

    Regular pay is the gross pay.

    Communicate: See completed payroll register on page 193.

3. **To Do:** Find Joseph Nitti's gross pay. Write the total hours, regular pay, and gross pay in the payroll register.

    Steps: 1. Find the total number of hours worked.

    2. Find the regular pay (gross pay).

    Information: Regular Pay Rate: $13 per hour.

    Hours Worked:

Monday	9 hours
Tuesday	8 hours
Wednesday	7 hours
Thursday	8 hours
Friday	8 hours

System of Measurement: Time and money

Computation:

$$9 + 8 + 7 + 8 + 8 = 40 \text{ hours}$$
$$40 \times \$13.00 = \$520$$

Regular pay is the gross pay.

Communicate: See completed payroll register on page 193.

4. **To Do:** Find Grace Pitini's gross pay. Write the total hours, regular pay, and gross pay in the payroll register.

    Steps: 1. Find the total number of hours worked.

            2. Find the regular pay (gross pay).

Information: Regular Pay Rate: $12.45 per hour.

Hours Worked:

Monday	7 hours
Tuesday	7 hours
Wednesday	7 hours
Thursday	7½ hours
Friday	7 hours

System of measurement: Time and money

Computation:

$$7 + 7 + 7 + 7\tfrac{1}{2} + 7 = 35\tfrac{1}{2} \text{ hours}$$
$$35\tfrac{1}{2} \times \$12.45 = \$441.98$$

Regular pay is the gross pay.

Communicate: See completed payroll register on page 193.

TASK 2

5. **To Do:** Find Ben Chu's gross pay. Write the total hours, regular pay, overtime pay, and gross pay in the payroll register.

    Steps: 1. Find the total number of hours and overtime hours.

            2. Figure the regular pay for the first 40 hours.

            3. Figure the overtime rate.

            4. Figure the overtime pay.

            5. Add the regular pay and overtime pay.

Information:

Monday	9 hours
Tuesday	9 hours
Wednesday	9 hours
Thursday	9 hours
Friday	9 hours

Regular rate: $6.70

System of Measurement: Time and money

Computation:

Total time: $9 + 9 + 9 + 9 + 9 = 45$ hours
Overtime hours: $45 - 40 = 5$
Regular pay: $40 \times \$6.70 = \$268$
Overtime rate: $\$6.70 \times 1.5 = \$10.05$
Overtime pay: $5 \times \$10.05 = \$50.25$
Gross pay: $\$268 + \$50.25 = \$318.25$

Communicate: See completed payroll register on page 193.

6. **To Do:** Find Cindy Masillo's gross pay. Write the total hours, regular pay, overtime pay, and gross pay in the payroll register.

    Steps: 1. Find the total number of hours and overtime hours.

               2. Figure the regular pay for the first 40 hours.

               3. Figure the overtime rate.

               4. Figure the overtime pay.

               5. Add the regular pay and overtime pay.

Information:

Monday	9 hours
Tuesday	10 hours
Wednesday	10 hours
Thursday	10 hours
Friday	8 hours

Regular rate: $5.75

System of Measurement: Time and money

Computation:

Total time: $9 + 10 + 10 + 10 + 8 = 47$ hours
Overtime hours: $47 - 40 = 7$
Regular pay: $40 \times \$5.75 = \$230$
Overtime rate: $\$5.75 \times 1.5 = \$8.63$
Overtime pay: $7 \times \$8.63 = \$60.41$
Gross pay: $\$230 + \$60.41 = \$290.41$

Communicate: See completed payroll register on page 193.

7. **To Do:** Find Janet Stein's gross pay. Write the total hours, regular pay, overtime pay, and gross pay in the payroll register.

    Steps: 1. Find the total number of hours and overtime hours.

               2. Figure the regular pay for the first 40 hours.

               3. Figure the overtime rate.

               4. Figure the overtime pay.

               5. Add the regular pay and overtime pay.

Information:

Monday	7½ hours
Tuesday	8 hours
Wednesday	8½ hours
Thursday	8 hours
Friday	8½ hours

Regular rate: $9.60

System of Measurement: Time and money

Computation:

Total time: $7.5 + 8 + 8.5 + 8 + 8.5 = 40.5$ hours
Overtime hours: $40.5 - 40 = .5$
Regular pay: $40 \times \$9.60 = \$384$
Overtime rate: $\$9.60 \times 1.5 = \$14.40$
Overtime pay: $.5 \times \$14.40 = \$7.20$
Gross pay: $\$384 + \$7.20 = \$391.20$

Communicate: See completed payroll register on page 193.

PAYROLL REGISTER   Week Ending 9/26

| EMPLOYEE INFORMATION | | | | GROSS EARNINGS | | | | DEDUCTIONS | | |
NAME	MAR STAT	ALLOW	TOTAL HOURS	REG RATE	REG PAY	OT PAY	GROSS PAY	FICA	FWT	NET PAY
M. Bayard	M	4	40	8.90	356.00		356.00			
S. Chester	S	1	43.5	8.45	338.00	44.38	382.38			
B. Chu	M	5	45	6.70	268.00	50.25	318.25			
B. Lopez	S	1	40	8.45	338.00		338.00			
S. Marino	M	6	36	8.40	302.40		302.40			
C. Masillo	S	3	47	5.75	230.00	60.41	290.41			
J. Nitti	M	2	40	13.00	520.00		520.00			
G. Pitini	M	0	35.5	12.45	441.98		441.98			
J. Stein	M	3	40.5	9.60	384.00	7.20	391.20			

TASK 3

8. **To Do:** Find the FICA tax for Sylvia Chester.

   Step: Multiply gross pay by 7.65%.

   Information: Gross pay is $382.38.

   System of Measurement: Money

   Computation: $382.38 \times 0.0765 = \$29.25$

   Communicate: See completed payroll register below.

9. **To Do:** Find the FICA tax for Ben Chu.

   Step: Multiply gross pay by 7.65%.

   Information: Gross pay is $318.25.

   System of Measurement: Money

   Computation: $318.25 \times 0.0765 = \$24.35$

   Communicate: See completed payroll register below.

10. **To Do:** Find the FICA tax for Barbara Lopez.

    Step: Multiply gross pay by 7.65%.

    Information: Gross pay is $338.00.

    System of Measurement: Money

    Computation: $338.00 \times 0.0765 = \$25.86$

    Communicate: See completed payroll register below.

PAYROLL REGISTER   Week Ending 9/26

| EMPLOYEE INFORMATION | | | | GROSS EARNINGS | | | | DEDUCTIONS | | |
NAME	MAR STAT	ALLOW	TOTAL HOURS	REG RATE	REG PAY	OT PAY	GROSS PAY	FICA	FWT	NET PAY
M. Bayard	M	4	40	8.90	356.00		356.00	27.23		
S. Chester	S	1	43.5	8.45	338.00	44.38	382.38	29.25		
B. Chu	M	5	45	6.70	268.00	50.25	318.25	24.35		
B. Lopez	S	1	40	8.45	338.00		338.00	25.86		
S. Marino	M	6	36	8.40	302.40		302.40			
C. Masillo	S	3	47	5.75	230.00	60.41	290.41			
J. Nitti	M	2	40	13.00	520.00		520.00			
G. Pitini	M	0	35.5	12.45	441.98		441.98			
J. Stein	M	3	40.5	9.60	384.00	7.20	391.20			

TASK 4

11. **To Do:** Figure the FWT and take-home pay for Sylvia Chester, and record them in the payroll register.

    Steps: 1. Find the FWT in the tax tables.

              2. Subtract FICA and FWT from the gross pay.

    Information: Gross pay $382.38; single; 1 allowance; FICA $29.25

    System of Measurement: Money

    Computation:

    From the Single Persons Weekly Payroll Period, 1 allowance = $48

$$\$382.38 - (\$29.25 + \$48) = \$305.13$$

    Communicate: See completed payroll register on page 195.

12. **To Do:** Figure the FWT and take-home pay for Ben Chu, and record them in the payroll register.

    Steps: 1. Find the FWT in the tax tables.

              2. Subtract FICA and FWT from the gross pay.

    Information: Gross pay $318.25; married; 5 allowances; FICA $24.35

    System of Measurement: Money

    Computation:

    From the Married Persons Weekly Payroll Period: $8

$$\$318.25 - (\$24.35 + \$8) = \$285.90$$

    Communicate: See completed payroll register on page 195.

13. **To Do:** Figure the FWT and take-home pay for Barbara Lopez, and record them in the payroll register.

    Steps: 1. Find the FWT in the tax tables.

              2. Subtract FICA and FWT from the gross pay.

    Information: Gross pay $338.00; single; 1 allowance; FICA $25.86

    System of Measurement: Money

    Computation:

$$FWT = \$41$$
$$\$338.00 - (\$25.86 + \$41) = \$271.14$$

    Communicate: See completed payroll register on page 195.

14. **To Do:** Figure the FWT and take-home pay for Susan Marino, and record them in the payroll register.

    Steps: 1. Find the FWT in the tax tables.

              2. Subtract FICA and FWT from the gross pay.

    Information: Gross pay $302.40; married; 6 allowances; FICA $23.13.

    System of Measurement: Money

    Computation:

$$FWT = \$0$$
$$\$302.40 - (\$23.13 + \$0) = \$279.27$$

    Communicate: See completed payroll register on page 195.

15. **To Do:** Figure the FWT and take-home pay for Cindy Masillo, and record them in the payroll register.

    Steps: 1. Find the FWT in the tax tables.

              2. Subtract FICA and FWT from the gross pay.

Information: Gross pay $290.41; single; 3 allowances; FICA $22.22

System of Measurement: Money

Computation:

$$FWT = \$23$$

$$\$290.41 - (\$22.22 + \$23) = \$245.19$$

Communicate: See completed payroll register below.

PAYROLL REGISTER   Week Ending 9/26

	EMPLOYEE INFORMATION			**GROSS EARNINGS**				**DEDUCTIONS**		
NAME	MAR STAT	ALLOW	TOTAL HOURS	REG RATE	REG PAY	OT PAY	GROSS PAY	FICA	FWT	NET PAY
M. Bayard	M	4	40	8.90	356.00		356.00	27.23	20.00	308.77
S. Chester	S	1	43.5	8.45	338.00	44.38	382.38	29.25	48.00	305.13
B. Chu	M	5	45	6.70	268.00	50.25	318.25	24.35	8.00	285.90
B. Lopez	S	1	40	8.45	338.00		338.00	25.86	41.00	271.14
S. Marino	M	6	36	8.40	302.40		302.40	23.13	0	279.27
C. Masillo	S	3	47	5.75	230.00	60.41	290.41	22.22	23.00	245.19
J. Nitti	M	2	40	13.00	520.00		520.00	39.78		
G. Pitini	M	0	35.5	12.45	441.98		441.98	33.81		
J. Stein	M	3	40.5	9.60	384.00	7.20	391.20	29.93		

TASK 5

16. **To Do:** Find the FICA tax, federal withholding tax, and net pay for Jack Brown. Record them in the payroll register.

    Steps: 1. Find the FICA tax by multiplying gross pay by 7.65%.

    2. Find the FWT in the tax tables.

    3. Find the net pay.

    Information: Gross pay $1,625.00; single; 0 allowances.

    System of Measurement: Money

    Computation:

    FICA: $1,625.00 × 7.65% = $124.31

    FWT: $337

    Net Pay: $1,625.00 − ($124.31 + $337) = $1,163.69

    Communicate: See completed payroll register on page 196.

17. **To Do:** Find the FICA tax, federal withholding tax, and net pay for Anna Hernandez. Record them in the payroll register.

    Steps: 1. Find the FICA tax by multiplying gross pay by 7.65%.

    2. Find the FWT in the tax tables.

    3. Find the net pay.

    Information: Gross pay $985.00; married; 4 allowances.

    System of Measurement: Money

    Computation:

    FICA: $985.00 × 7.65% = $75.35

    FWT: $76

    Net Pay: $985.00 − ($75.355 + $76) = $833.65

    Communicate: See completed payroll register on page 196.

18. **To Do:** Find the FICA tax, federal withholding tax, and net pay for Sara Jackson. Record them in the payroll register.

   Steps: 1. Find the FICA tax by multiplying gross pay by 7.65%.

   2. Find the FWT in the tax tables.

   3. Find the net pay.

   Information: Gross pay $1,166.66; married; 3 allowances.

   System of Measurement: Money

   Computation:

   FICA: $1,166.66 × 7.65% = $89.25

   FWT: $116

   Net Pay: $1,166.66 − ($89.25 + $116) = $961.41

   Communicate: See completed payroll register below.

19. **To Do:** Find the FICA tax, federal withholding tax, and net pay for Barbara Mendez. Record them in the payroll register.

   Steps: 1. Find the FICA tax by multiplying gross pay by 7.65%.

   2. Find the FWT in the tax tables.

   3. Find the net pay.

   Information: Gross pay $1,333.33; single; 1 allowance.

   System of Measurement: Money

   Computation:

   FICA: $1,333.33 × 7.65% = $102.00

   FWT: $229

   Net Pay: $1,333.33 − ($102.00 + $229) = $1,002.33

   Communicate: See completed payroll register below.

PAYROLL REGISTER   Week Ending 9/26

EMPLOYEE INFORMATION			GROSS EARNINGS	DEDUCTIONS		
NAME	MAR STAT	ALLOW	GROSS PAY	FICA	FWT	NET PAY
M. Anderson	M	2	1208.33	92.44	135.00	980.89
J. Brown	S	0	1625.00	124.31	337.00	1163.69
A. Hernandez	M	4	985.00	75.35	76.00	833.65
S. Jackson	M	3	1166.66	89.25	116.00	961.41
B. Mendez	S	1	1333.33	102.00	229.00	1002.33

Answer Key

JOB SITUATION 1

TASK 1

1. **To Do:** Find out whether to send first class or Priority and how much postage is needed for each letter.

   Steps: 1. Decide whether to use first-class mail or Priority Mail.

   2. Look up the rate in the postal rate tables.

   Information: The letters weigh 2½ ounces, 3 ounces, and 8 ounces.

   System of Measurement: Weight and money

   Computation:

   Send First Class: 2½ ounces = 3-ounce rate = $0.75

   Send First Class: 3-ounce rate = $0.75

   Send First Class: 8-ounce rate = $1.90

   Communicate:

   $0.75 First Class

   $0.75 First Class

   $1.90 First Class

2. **To Do:** Find out whether to send first class or Priority and how much postage is needed for each letter.

   Steps: 1. Decide whether to use first-class mail or Priority Mail.

   2. Look up the rate in the postal rate tables.

   Information: The letters weigh 5 ounces, 1 pound, and 2 ounces.

   System of Measurement: Weight and money

   Computation:

   Send First Class: 5-ounce rate = $1.21

   Send Priority: 1-pound rate = $2.90

   Send First Class: 2-ounce rate = $0.52

   Communicate:

   $1.21 First Class

   $2.90 Priority Mail

   $0.52 First Class

3. **To Do:** Find out whether to send first class or Priority and how much postage is needed for each letter.

   Steps: 1. Decide whether to use first-class mail or Priority Mail.

   2. Look up the rate in the postal rate tables.

   Information: The letters weigh 3½ ounces, 1¼ pounds, 1 ounce, and 4 ounces.

   System of Measurement: Weight and money

Computation:

> Send First Class: 3½ ounces = 4-ounce rate = $0.98
>
> Send Priority: 1¼ pounds = 2-pound rate = $2.90
>
> Send First Class: 1-ounce rate = $0.29
>
> Send First Class: 4-ounce rate = $0.98

Communicate:

$0.98 First Class

$2.90 Priority Mail

$0.29 First Class

$0.98 First Class

4. **To Do:** Find out whether to send first class or Priority and how much postage is needed for each letter.

   Steps: 1. Decide whether to use first-class mail or Priority Mail.

   2. Look up the rate in the postal rate tables.

   Information: The envelopes weigh 10½ ounces, 3½ pounds, 7¼ ounces, and just a shade over 11 ounces.

   System of Measurement: Weight and money

   Computation:

   > Send First Class: 10½ ounces = 11-ounce rate = $2.59
   >
   > Send Priority: 3½ pounds = 4-pound rate = $4.65
   >
   > Send First Class: 7¼ ounces = 8-ounce rate = $1.90
   >
   > Send Priority: over 11-ounce rate = $2.90

   Communicate:

   $2.59 First Class

   $4.65 Priority Mail

   $1.90 First Class

   $2.90 Priority Mail

5. **To Do:** Find out whether to send first class or Priority and how much postage is needed for each letter.

   Steps: 1. Decide whether to use first-class mail or Priority Mail.

   2. Look up the rate in the postal rate tables.

   Information: The envelopes weigh under 1 ounce, 6½ ounces, 5 pounds, and 17 ounces.

   System of Measurement: Weight and money

   Computation:

   > Send First Class: under 1 ounce = 1-ounce rate = $0.29
   >
   > Send First Class: 6½ ounces = 7-ounce rate = $1.67
   >
   > Send Priority: 5-pound rate = $5.45
   >
   > Send Priority: 17 ounces = over 11-ounce rate = $2.90

   Communicate:

   > $0.29 First Class
   >
   > $1.67 First Class
   >
   > $5.45 Priority Mail
   >
   > $2.90 Priority Mail

TASK 2

6. **To Do:** Find how much postage is needed.
   Step: Find the rate for the weight and zone.
   Information: Weight = 41 pounds; Zone 4
   System of Measurement: Weight and money
   Computation:
   > Look in Zone 4.
   > Rate for 41 pounds = $7.26
   Communicate:
   > $7.26 Parcel Post, or Fourth Class

7. **To Do:** Find how much postage is needed.
   Step: Find the rate for the weight and zone.
   Information: Weight = 8 pounds 12 ounces; Zone 5
   System of Measurement: Weight and money
   Computation:
   > Look in Zone 5.
   > Rate for 8 pounds 12 ounces = rate for 9 pounds = $5.40
   Communicate:
   $4.60 Parcel Post, or Fourth Class

8. **To Do:** Find how much postage is needed.
   Step: Find the rate for the weight and zone.
   Information: Weight = 4 pounds 2 ounces; Zone 1
   System of Measurement: Weight and money
   Computation:
   > Look in Zone 1.
   > Rate for 4 pounds 2 ounces = rate for 5 pounds = $2.49
   Communicate:
   > $2.49 Parcel Post, or Fourth Class

9. **To Do:** Find how much postage is needed.
   Step: Find the rate for the weight and zone.
   Information: Weight = 50 pounds 8 ounces; Zone 8
   System of Measurement: Weight and money
   Computation:
   > Look in Zone 8.
   > Rate for 50 pounds 8 ounces = rate for 51 pounds = $29.33
   Communicate:
   > $29.33 Parcel Post, or Fourth Class

TASK 3

10. **To Do:** Find the rate for sending by UPS.
    Steps: 1. Find the zone for the ZIP Code.
    2. Find the rate in the Residential Deliveries Rate Chart.

Information: Home address; ZIP Code starts with 577; weight 22 pounds

System of Measurement: Weight and money

Computation:

        Zone is 7 for 577...

        Rate for 22 pounds = $9.48

Communicate:

        $9.48 UPS

11. **To Do:** Find the rate for sending by UPS.

    Steps: 1. Find the zone for the ZIP Code.

               2. Find the rate in the Commercial Deliveries Rate Chart.

    Information: Business address; ZIP Code starts with 100; weight 15 pounds 11 ounces

    System of Measurement: Weight and money

    Computation:

        Zone is 2 for 100..

        Rate for 15 pounds 11 ounces = rate for 16 pounds = $3.03

    Communicate:

        $3.03 UPS

12. **To Do:** Find the rate for sending by UPS.

    Steps: 1. Find the zone for the ZIP Code.

               2. Find the rate in the Commercial Deliveries Rate Chart.

    Information: Business address; ZIP Code starts with 975; weight 41 pounds 1 ounce

    System of Measurement: Weight and money

    Computation:

        Zone is 8 for 975..

        Rate for 41 pounds 1 ounce = rate for 42 pounds = $19.64

    Communicate:

        $19.64 UPS

TASK 4

13. **To Do:** Find the best shipping method and the shipping cost.

    Steps: 1. Find the fourth-class rate.

               2. Find the UPS rate.

               3. Compare the rates and choose fourth class or UPS.

    Information: Business address; ZIP Code starts with 338; U.S.P.S. Zone 5; weight 69 pounds

    System of Measurement: Weight and money

    Computation:

        1. 69 pounds in Zone 5 = $11.98 Fourth Class

        2. Zone = 5 UPS

69 pounds in Zone 5 = $13.64 (commercial rate)

   3. Subtract $13.64 – $11.98 = $1.66; greater than $1.00; choose fourth class

Communicate:

      $11.98 Fourth Class

14. **To Do:** Find the best shipping method and the shipping cost.

   Steps: 1. Find the fourth-class rate.

         2. Find the UPS rate.

         3. Compare the rates and choose fourth class or UPS.

   Information: Business address; ZIP Code starts with 881; U.S.P.S. Zone 7; weight 10 pound 14 ounces

   System of Measurement: Weight and money

   Computation:

      1. 11 pounds in Zone 7 = $9.97 Fourth Class

      2. Zone = 7 UPS

        11 pounds in Zone 7 = $5.27 (commercial rate)

      3. UPS rate is less. Choose UPS.

   Communicate:

      $5.27 UPS

15. **To Do:** Find the best shipping method and the shipping cost.

   Steps: 1. Find the fourth-class rate.

         2. Find the UPS rate.

         3. Compare the rates and choose fourth class or UPS.

   Information: Business address; ZIP Code starts with 778; U.S.P.S. Zone 6; weight 40 pounds

   System of Measurement: Weight and money

   Computation:

      1. 40 pounds in Zone 6 = $12.92 Fourth Class

      2. Zone = 7 UPS

        40 pounds in Zone 7 = $15.81 (commercial rate)

      3. Subtract $15.81 – $12.92 = $2.89; greater than $1.00; choose fourth class

   Communicate:

      $12.92 Fourth Class

16. **To Do:** Find the best shipping method and the shipping cost.

   Steps: 1. Find the fourth-class rate.

         2. Find the UPS rate.

         3. Compare the rates and choose fourth class or UPS.

   Information: Business address; ZIP Code starts with 475; U.S.P.S. Zone 5; weight 45 pounds

   System of Measurement: Weight and money

   Computation:

      1. 45 pounds in Zone 5 = $10.46 Fourth Class

      2. Zone = 5 UPS

45 pounds in Zone 5 = $11.42 (commercial rate)

   3. Subtract $11.42 – $10.46 = $0.96; less than $1.00; choose UPS

Communicate:

   $11.42 UPS

TASK 5

17. **To Do:** Pick the type of delivery service and find the cost.

   Steps: 1. Choose the latest possible delivery.

            2. Find the rate.

   Information: Delivery by 5 p.m.; weight 3 pounds

   System of Measurement: Weight and money

   Computation:

        To get there by 5:00, use Next Day Service.

        Next Day Service rate for 3 pounds is $17.00.

   Communicate:

        $17.00 Next Day Service

18. **To Do:** Pick the type of delivery service and find the cost.

   Steps: 1. Choose the latest possible delivery.

            2. Find the rate.

   Information: Delivery by 4 p.m. the day after tomorrow; weight 8 ounces

   System of Measurement: Weight and money

   Computation:

        To get there by 4:00 the day after tomorrow, use Second Day Service.

        Second Day Service rate for 8 ounces is $7.95.

   Communicate:

        $7.95 Second Day Service

19. **To Do:** Pick the type of delivery service and find the cost.

   Steps: 1. Choose the latest possible delivery.

            2. Find the rate.

   Information: Delivery by 1 p.m. tomorrow; weight 6 pounds

   System of Measurement: Weight and money

   Computation:

        To get there by 1:00 tomorrow, use Priority Service.

        Priority Service rate for 6 pounds is $22.50.

   Communicate:

        $22.50 Priority Service

JOB SITUATION 2

TASK 1

1. **To Do:** Find the percent to use.

   Step: Look up the original and copy size in the chart.

   Information: The original is 11 by 15 inches. The copy is to be 8½ by 11 inches.

   System of Measurement: Length and width

   Computation: The chart shows that the reduction is 74%.

   Communicate: 74% reduction

2. **To Do:** Find the percent to use.

   Step: Look up the original and copy size in the chart.

   Information: The original is 8½ by 11 inches. The copy is to be 11 by 17 inches.

   System of Measurement: Length and width

   Computation: The chart shows that the enlargement is 129%.

   Communicate: 129% enlargement

TASK 2

3. **To Do:** Find the percent to use and the new length.

   Steps: 1. Solve an equation to find the percent.
   2. Solve a proportion to find the length.

   Information:

   original width = 6 inches

   original length = 8 inches

   new width = 3 inches

   System of Measurement: Length and width

   Computation:

   $$n \times 6 = 3$$
   $$\div 6 \qquad \div 6$$
   $$n = .50$$
   $$50\%$$

   $$\frac{6}{8} = \frac{3}{n} \qquad 6 \times n = 24$$
   $$\div 6 \qquad \div 6$$
   $$n = 4 \text{ inches}$$

   Communicate: 50% 4 inches

4. **To Do:** Find the percent to use and the new length.

   Steps: 1. Solve an equation to find the percent.
   2. Solve a proportion to find the length.

   Information:

   original width = 6 inches

   original length = 9 inches

   new width = 4½ inches

   System of Measurement: Length and width

Computation:

$$n \times 6 = 4\frac{1}{2}$$
$$n = .75$$
$$75\%$$

$$\frac{6}{9} = \frac{4.5}{n} \qquad n = 6.75 \text{ inches}$$

Communicate: 75% 6.75 inches

5. **To Do:** Find the percent to use and the new length.

    Steps: 1. Solve an equation to find the percent.

             2. Solve a proportion to find the length.

    Information:

        original width = 8 inches

        original length = 10 inches

        new width = 2¾ inches

    System of Measurement: Length and width

    Computation:

$$n \times 8 = 2\frac{3}{4}$$
$$n = .34$$
$$34\%$$

$$\frac{8}{10} = \frac{2.75}{n} \qquad n = 3.4 \text{ inches}$$

Communicate: 34% 3.4 inches

6. **To Do:** Find the percent to use and the new length.

    Steps: 1. Solve an equation to find the percent.

             2. Solve a proportion to find the length.

    Information:

        original width = 5 inches

        original length = 7 inches

        new width = 8 inches

    System of Measurement: Length and width

    Computation:

$$n \times 5 = 8$$
$$n = 1.6$$
$$160\%$$

$$\frac{5}{7} = \frac{8}{n} \qquad n = 11.2 \text{ inches}$$

Communicate: 160% 11.2 inches

7. **To Do:** Find the percent to use and the new length.

    Steps: 1. Solve an equation to find the percent.

             2. Solve a proportion to find the length.

    Information:

        original width = 10½ inches

        original length = 12 inches

        new width = 15¾ inches

System of Measurement: Length and width

Computation:

$$n \times 10\tfrac{1}{2} = 15\tfrac{3}{4}$$
$$n = 1.5$$
$$150\%$$

$$\frac{10.5}{12} = \frac{15.75}{n} \qquad n = 18 \text{ inches}$$

Communicate: 150% 18 inches

TASK 3

8. **To Do:** Find the cost of the copies.

   Steps: 1. Find the total number of copies.

   2. Multiply the rate times the total number of copies.

   3. Add the cost for any extra services.

   Information: Accounting Department, 10 copies, 5 pages each

   System of Measurement: Money

   Computation:

   10 copies × 5 pages = 10 × 5 = 50

   50 copies × 7¢ a copy = 50 × 0.07 = $3.50

   Communication: See the completed Copy Log on page 206.

9. **To Do:** Find the cost of the copies.

   Steps: 1. Find the total number of copies.

   2. Multiply the rate times the total number of copies.

   3. Add the cost for any extra services.

   Information: Accounting Department, 20 copies, 14 pages each, legal size

   System of Measurement: Money

   Computation:

   20 × 14 = 280

   280 × 0.08 = $22.40

   Communicate: See the completed Copy Log on page 206.

10. **To Do:** Find the cost of the copies.

    Steps: 1. Find the total number of copies.

    2. Multiply the rate times the total number of copies.

    3. Add the cost for any extra services.

    Information: Sales Department, 30 copies, 27 pages each, flat comb binding

    System of Measurement: Money

    Computation:

    30 × 27 = 810

    810 × 0.07 = $56.70

    flat comb binding: 30 × .10 = $3.00

    total: $56.70 + $3.00 = $59.70

    Communicate: See the completed Copy Log on page 206.

11. **To Do:** Find the cost of the copies.

    Steps: 1. Find the total number of copies.

          2. Multiply the rate times the total number of copies.

          3. Add the cost for any extra services.

    Information: Sales Department, 200 copies, 30 pages each, offset duplication

    System of Measurement: Money

    Computation:

$$200 \times 30 = 6,000$$
$$6,000 \times 0.025 = \$150$$

    Communicate: See the completed Copy Log below.

12. **To Do:** Find the cost of the copies.

    Steps: 1. Find the total number of copies.

          2. Multiply the rate times the total number of copies.

          3. Add the cost for any extra services.

    Information: Data Processing Department, 100 copies, 25 pages each, offset duplication, spiral comb binding

    System of Measurement: Money

    Computation:

$$100 \times 25 = 2,500$$
$$2,500 \times 0.045 = \$112.50$$
$$\text{spiral comb binding: } 100 \times .25 = \$25$$
$$\text{total: } \$112.50 + \$25.00 = 137.50$$

    Communicate: See the completed Copy Log below.

Date: September 24

Department	Total Copies	Special Services	Total Cost of Copies
DP	350	spiral bind 50 copies	$37.00
Acct.	50		$3.50
Acct.	280		$22.40
Sales	810	flat bind 30 copies	$59.70
Sales	6,000	offset	$150.00
DP	2,500	offset spiral bind 100 copies	$137.50

JOB SITUATION 3

TASK 1

1. **To Do:** Pick the flight for this trip.

Steps: 1. Look for a flight at or after the time your supervisor wants to fly.

2. Check to be sure the flight operates on the day you want.

Information: Monday, from Chicago to Dallas, around 3:00 p.m.

System of Measurement: Time

Computation:

> *OAG* shows flight at 3:05 p.m., arriving 5:40 p.m., all days, American Airlines, flight number 121.

Communicate:

> Monday: Chicago to Dallas, depart 3:05 p.m., arrive 5:40 p.m., American Airlines 121.

2. **To Do:** Pick the flight for this trip.

Steps: 1. Look for a flight at or after the time your supervisor wants to fly.

2. Check to be sure the flight operates on the day you want.

Information: Tuesday, from Seattle to Kansas City, after noon

System of Measurement: Time

Computation:

> *OAG* shows flight at 12:45 p.m., arriving 5:55 p.m., all days, Northwest Airlines, flight number 66.

Communicate:

> Tuesday: Seattle to Kansas City, depart 12:45 p.m., arrive 5:55 p.m., Northwest Airlines 66.

3. **To Do:** Pick the flight for this trip.

Steps: 1. Look for a flight at or after the time your supervisor wants to fly.

2. Check to be sure the flight operates on the day you want.

Information: Wednesday, from Chicago to Denver, around 11 a.m.

System of Measurement: Time

Computation:

> *OAG* shows flight at 11 a.m., arriving 12:18 p.m., all days, United Airlines, flight number 227.

Communicate:

> Wednesday: Chicago to Denver, depart 11 a.m., arrive 12:18 p.m., United Airlines 227.

4. **To Do:** Pick the flights for this trip.

Steps: 1. Look for flights at or after the time your supervisor wants to fly.

2. Check to be sure the flights operate on the day you want.

Information: Thursday, from Chicago to Madison, 7 a.m., return 7 p.m.

System of Measurement: Time

Computation:

> *OAG* shows flight at 7 a.m., arriving 7:55 a.m., not Saturday or Sunday, American Airlines, flight number 4221;
>
> Flight at 8:25 p.m., arriving 9:05 p.m., not Saturday, United Airlines, flight number 2858.

Communicate:

> Thursday: Chicago to Madison, depart 7 a.m., arrive 7:55 p.m., American Airlines 4221.
>
> Thursday: Madison to Chicago, depart 8:25 p.m., arrive 9:05 p.m., American Airlines flight 2858.

5. **To Do:** Pick the flight for this trip.

   Steps: 1. Look for a flight at or after the time your supervisor wants to fly.

            2. Check to be sure the flight operates on the day you want.

Information: Tuesday from Chicago to Missoula, Montana, after 4:30 p.m. and return to Chicago around noon on Friday

System of Measurement: Time

Computation:

> *OAG* shows connecting flight Northwest Airlines flight 507 at 4:40 p.m., arriving Minneapolis at 6:09 p.m; flight 709 leaving 7 p.m. and arriving Missoula at 9:25 p.m.
>
> Return connecting flight Northwest Airlines flight 708 leaving 12:25 p.m., arriving Minneapolis 4:50 p.m.; flight 784 leaving 6:00 p.m. and arriving 7:17 p.m.
>
> Both connecting flights operate all days.

Communicate:

Tuesday: Chicago to Missoula via Minneapolis:

> ❏ Chicago to Minneapolis, depart 4:40 p.m., arrive 6:09 p.m., Northwest Airlines 507.
>
> ❏ Minneapolis to Missoula, depart 7 p.m., arrive 9:25 p.m., Northwest Airlines 709.

 Friday: Missoula to Chicago via Minneapolis

> ❏ Missoula to Minneapolis, depart 12:25 p.m., arrive 4:50 p.m., Northwest Airlines 708.
>
> ❏ Minneapolis to Chicago, depart 6 p.m., arrive 7:17 p.m., Northwest Airlines 784.

TASK 2

6. **To Do:** Estimate the total car rental costs.

   Steps: 1. Multiply the number of days times the correct daily rate.

            2. Multiply the miles times the mileage fee.

            3. Multiply the daily insurance cost times the number of days.

            4. Add these three fees to get the total fee.

Information: Luxury car, one day (weekday), 100 miles

System of Measurement: Money, time, and distance

Computation:

   Daily rate $49.95 × 1 day = $49.95

   Mileage rate $0.18 per mile × 100 miles = $18.00

   Insurance rate $9.00 per day × 1 days = $9.00

   Total: $49.95 + $18.00 + $9.00 = $76.95

Communicate: $73.95

7. **To Do:** Estimate the total car rental costs.

   Steps: 1. Multiply the number of days times the correct daily rate.

   2. Multiply the miles times the mileage fee.

   3. Multiply the daily insurance cost times the number of days.

   4. Add these three fees to get the total fee.

   Information: Van, 2 days (Tuesday and Wednesday), 700 miles

   System of Measurement: Money, time, and distance

   Computation:

   Daily rate $49.95 × 2 days = $99.90

   Mileage rate $0.21 per mile × 700 miles = $147.00

   Insurance rate $9.00 per day × 2 days = $18.00

   Total: $99.90 + $147.00 + $18.00 = $264.90

   Communicate: $258.90

8. **To Do:** Estimate the total car rental costs.

   Steps: 1. Multiply the number of days times the correct daily rate.

   2. Multiply the miles times the mileage fee.

   3. Multiply the daily insurance cost times the number of days.

   4. Add these three fees to get the total fee.

   Information: Four-door sedan, 2 days (1 weekday and 1 weekend), 200 miles

   System of Measurement: Money, time, and distance

   Computation:

   Daily rate weekday: $35.95 × 1 day = $35.95

   Daily rate weekend: $24.95 × 1 day = $24.95

   Mileage rate $0.16 per mile × 200 miles = $32.00

   Insurance rate $6.00 per day × 2 days = $12.00

   Total: $35.95 + $24.95 + $32.00 + $12.00 = $104.90

   Communicate: $104.90

9. **To Do:** Estimate the total car rental costs.

   Steps: 1. Multiply the number of days times the correct daily rate.

   2. Multiply the miles times the mileage fee.

   3. Multiply the daily insurance cost times the number of days.

   4. Add these three fees to get the total fee.

   Information: Subcompact, 1 week (5 weekdays and 2 weekend days), 400 miles

   System of Measurement: Money, time, and distance

Computation:

        Daily rate weekday: $24.95 × 5 days = $124.75

        Daily rate weekend: $19.99 × 2 days = $39.98

        Mileage rate $0.16 per mile × 400 miles = $64.00

        Insurance rate $6.00 per day × 7 days = $42.00

        Total: $124.75 + $39.98 + $64.00 + $42.00 = $270.73

Communicate: $270.73

10. **To Do:** Estimate the total car rental costs. Decide which is less expensive, the regular rate or the special rate.

    Steps: 1. Multiply the number of days times the correct daily rate.

          2. Multiply the miles times the mileage fee.

          3. Multiply the daily insurance cost times the number of days.

          4. Add these three fees to get the total fee.

          5. Compare this to the special rate.

    Information: Four-door sedan, 1 weekday, 100 miles; special rate $45.

    System of Measurement: Money, time, and distance

    Computation:

        Daily rate $35.95 × 1 days = $35.95

        Mileage rate $0.16 per mile × 100 miles = $16.00

        Insurance rate $6.00 per day × 1 days = $6.00

        Total: $35.95 + $16.00 + $6.00 = $57.95

    Communicate: Special rate; $45

11. **To Do:** Estimate the total car rental costs. Decide which is less expensive, the regular rate or the special rate.

    Steps: 1. Multiply the number of days times the correct daily rate.

          2. Multiply the miles times the mileage fee.

          3. Multiply the daily insurance cost times the number of days.

          4. Add these three fees to get the total fee.

          5. Compare this to the special rate.

    Information: Subcompact, 4 weekdays, 300 miles; special rate $150.

    System of Measurement: Money, time, and distance

    Computation:

        Daily rate $24.95 × 4 days = $99.80

        Mileage rate $0.16 per mile × 300 miles = $48.00

        Insurance rate $6.00 per day × 4 days = $24.00

        Total: $99.80 + $48.00 + $24.00 = $171.80

    Communicate: Special rate; $150

Answer Key

JOB SITUATION 1

TASK 1

1. **To Do:** Draw a bar graph to show sales per salesperson.

   Steps: 1. Create a table for the data.

   2. Choose a label for each axis.

   3. Write the data on each axis. Choose a scale and interval for the numerical axis.

   4. Draw a bar for each pair of numbers in your table.

   5. Title the graph.

   Information: Marcus sold $80,000 in 1992. Williams sold $40,000. Macintosh sold $90,000.

   System of Measurement: Money

   Computation:

   1. Table:

Salesperson	Sales
Marcus	$80,000
Williams	$40,000
Macintosh	$90,000

   2. Labels: "Salespeople"; "Sales in Thousands of Dollars"

   3. Scale: $0 to $100,000; Interval: $10,000

   4. Bars: See the graph below.

   5. Title: "1992 Sales"

   Communicate: See the graph below.

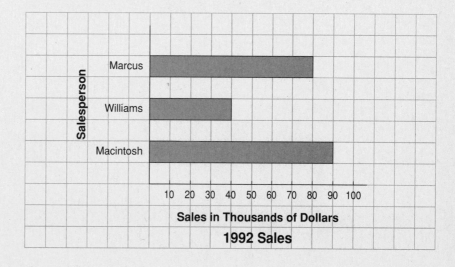

2. **To Do:** Draw a bar graph to show sales per group.

  Steps: 1. Create a table for the data.

  2. Choose a label for each axis.

  3. Write the data on each axis. Choose a scale and interval for the numerical axis.

  4. Draw a bar for each pair of numbers in your table.

  5. Title the graph.

Information: Group 1 sold $200,000 in 1991. Group 2 sold $100,000. Group 3 sold $150,000.

System of Measurement: Money

Computation:

  1. Table:

Group	Sales
Group 1	$200,000
Group 2	$100,000
Group 3	$150,000

  2. Labels: "Sales Groups"; "Sales in Thousands of Dollars"

  3. Scale: $250,000 to $200,000; Interval: $25,000 (Other choices may be acceptable as well.)

  4. Bars: See the graph below.

  5. Title: "1992 Sales"

Communicate: See the graph below.

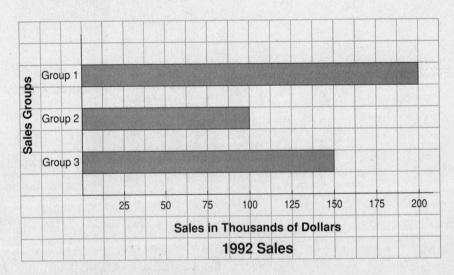

3. **To Do:** Draw a bar graph to show sales per salesperson.

  Steps: 1. Create a table for the data.

  2. Choose a label for each axis.

  3. Write the data on each axis. Choose a scale and interval for the numerical axis.

  4. Draw a bar for each pair of numbers in your table.

  5. Title the graph.

Information: Saunders sold $20,000 in July. Miller sold $15,000 in July. White sold $10,000 in July. Mendez sold $20,000 in July.

System of Measurement: Money

Computation:

    1. Table:

Salesperson	Sales
Saunders	$20,000
Miller	$15,000
White	$10,000
Mendez	$20,000

    2. Labels: "Salespeople"; "Sales in Thousands of Dollars"

    3. Scale: $0 to $25,000; Interval: $5,000 (Other choices may be acceptable as well.)

    4. Bars: See the graph below.

    5. Title: "July Sales"

Communicate: See the graph below.

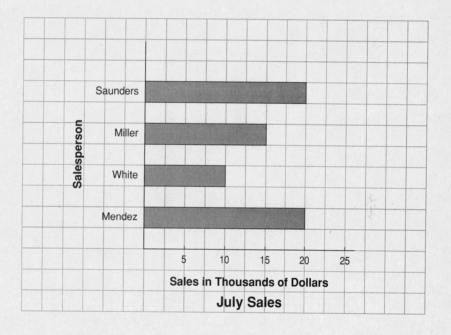

4. **To Do:** Draw a bar graph to show sales per year.

    Steps: 1. Create a table for the data.

            2. Choose a label for each axis.

            3. Write the data on each axis. Choose a scale and interval for the numerical axis.

            4. Draw a bar for each pair of numbers in your table.

            5. Title the graph.

Information: Sales were $1,000,000 in 1989; $1,100,000 in 1990; $1,200,000 in 1991; and $1,350,000 in 1992.

System of Measurement: Money

Computation:

    1. Table:

Year	Sales
1989	$1,000,000
1990	$1,100,000
1991	$1,200,000
1992	$1,350,000

    2. Labels: "Years"; "Sales in Million of Dollars"

    3. Scale: $1,000,000 to $1,500,000; Interval: $100,000 (Other choices may be acceptable as well.)

    4. Bars: See the graph below.

    5. Title: "Sales From 1989 to 1992"

Communicate: See the graph below.

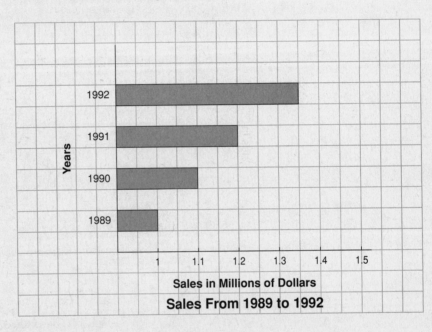

TASK 2

    5. **To Do:** Draw a line graph to show Anderson's sales per year.

    Steps: 1. Create a table for the data.

            2. Choose a label for each axis.

            3. Write the data on each axis. Choose a scale and interval for the numerical axis.

            4. Draw a point for each pair of numbers in your table. Connect the points.

            5. Title the graph.

    Information: Sales were $75,000 in 1989; $90,000 in 1990; $100,000 in 1991; and $120,000 in 1992.

    System of Measurement: Money

Computation:

1. Table:

Year	Sales
1989	$75,000
1990	$90,000
1991	$100,000
1992	$120,000

2. Labels: "Years"; "Sales in Thousands of Dollars"

3. Scale: $70,000 to $120,000; Interval: $10,000 (Other choices may be acceptable as well.)

4. Lines: See the graph below.

5. Title: "Anderson's Sales"

Communicate: See the graph below.

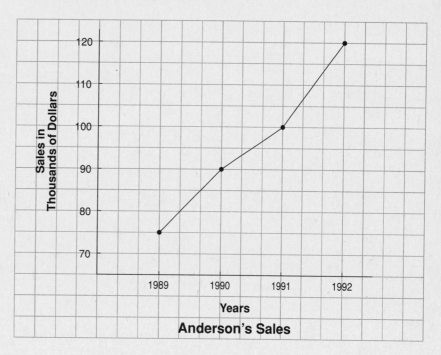

6. **To Do:** Draw a line graph to show Group 1's sales per year.

   Steps: 1. Create a table for the data.

   2. Choose a label for each axis.

   3. Write the data on each axis. Choose a scale and interval for the numerical axis.

   4. Draw a point for each pair of numbers in your table. Connect the points.

   5. Title the graph.

Information: Sales were $135,000 in 1989; $145,000 in 1990; $180,000 in 1991; and $200,000 in 1992.

System of Measurement: Money

Computation:

    1. Table:

Year	Sales
1989	$135,000
1990	$145,000
1991	$180,000
1992	$200,000

    2. Labels: "Years"; "Sales in Thousands of Dollars"

    3. Scale: $130,000 to $200,000; Interval: $10,000 (Other choices may be acceptable as well.)

    4. Lines: See the graph below.

    5. Title: "Group 1's Sales"

Communicate: See the graph below.

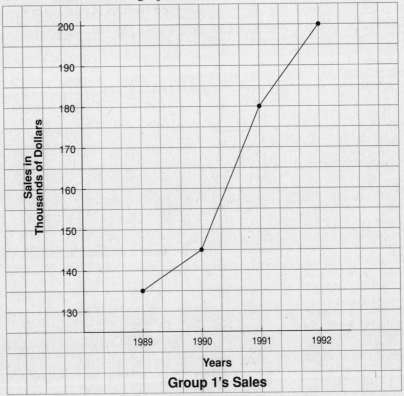

7. **To Do:** Draw a line graph to show Illinois sales for January to May.

   Steps: 1. Create a table for the data.

         2. Choose a label for each axis.

         3. Write the data on each axis. Choose a scale and interval for the numerical axis.

         4. Draw a point for each pair of numbers in your table. Connect the points.

         5. Title the graph.

Information: Sales were $50,000 in January; $34,500 in February; $40,000 in March; $60,000 in April; and $62,000 in May.

System of Measurement: Money

Computation:

    1. Table:

Month	Sales
January	$50,000
February	$34,500
March	$40,000
April	$60,000
May	$62,000

    2. Labels: "Months"; "Sales in Thousands of Dollars"

    3. Scale: $30,000 to $65,000; Interval: $5,000 (Other choices may be acceptable as well.)

    4. Lines: See the graph below.

    5. Title: "Illinois Sales From January to May"

Communicate: See the graph below.

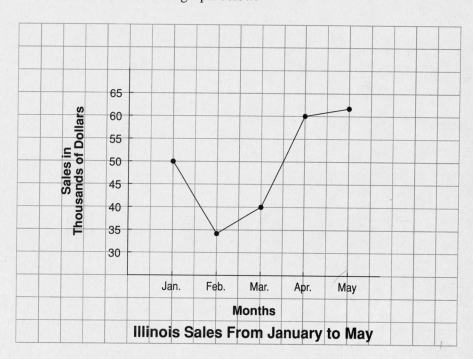

8. **To Do:** Draw a line graph to show sales in Iowa for June to October.

    Steps: 1. Create a table for the data.

             2. Choose a label for each axis.

             3. Write the data on each axis. Choose a scale and interval for the numerical axis.

             4. Draw a point for each pair of numbers in your table. Connect the points.

             5. Title the graph.

Information: Sales were $6,000 in June; $4,500 in July; $4,000 in August; $7,100 in September; and $7,000 in October.

System of Measurement: Money

Computation:

1. Table:

Month	Sales
June	$6,000
July	$4,500
August	$4,000
September	$7,100
October	$7,000

2. Labels: "Months"; "Sales in Dollars"

3. Scale: $4,000 to $7,500; Interval: $500 (Other choices may be acceptable as well.)

4. Lines: See the graph below.

5. Title: "Iowa Sales From June to October"

Communicate: See the graph below.

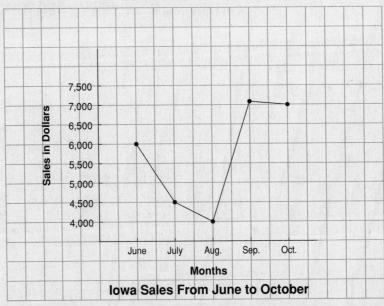

TASK 3

9. **To Do:** Draw a circle graph to show days not worked in the Accounting Department.

Steps: 1. Create a table for the data.

2. Find the percentage of the whole for each item.

3. Find the number of degrees for each item.

4. Draw a circle and draw an angle for each item of data. Label it.

5. Title the graph.

Information: Total days: 100. Causes: 50 vacation days; 25 sick days; 25 personal days.

System of Measurement: Time

Computation:

Cause	Vacation	Sick	Personal
Days	50	25	25

Vacation days: 50%   180°

Sick days: 25%   90°

Personal days: 25%   90°

Communicate:

**Days Not Worked in
Accounting Department**

10. **To Do:** Draw a circle graph to show days not worked in the Mailroom.

   Steps: 1. Create a table for the data.

   2. Find the percentage of the whole for each item.

   3. Find the number of degrees for each item.

   4. Draw a circle and draw an angle for each item of data. Label it.

   5. Title the graph.

Information: Total days: 200. Causes: 100 vacation days; 60 sick days; 40 personal days.

System of Measurement: Time

Computation:

Cause	Vacation	Sick	Personal
Days	100	60	40

Vacation days: 50%   180°

Sick days: 30%   108°

Personal days: 20%   72°

Communicate:

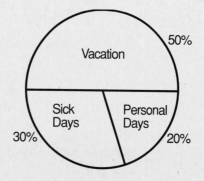

**Days Not Worked in
Mailroom**

11. **To Do:** Draw a circle graph to show the number of employees in each department.

   Steps: 1. Create a table for the data.

   2. Find the percentage of the whole for each item.

   3. Find the number of degrees for each item.

   4. Draw a circle and draw an angle for each item of data. Label it.

   5. Title the graph.

   Information: Total employees 200. Departments: data processing — 50; accounting — 20; mailroom — 10; sales — 70; office assistants — 50.

   System of Measurement: People

   Computation:

Dept.	DP	Acct.	Mail	Sales	OA
Emplys.	50	20	10	70	50

   Data Processing: 25%   90°

   Accounting: 10%   36°

   Mailroom: 5%   18°

   Sales: 35%   126°

   Office Assistants: 25%   90°

   Communicate:

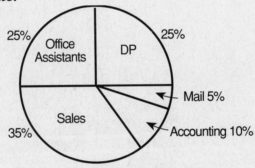

**Percentage of Employees
Per Department**

12. **To Do:** Draw a circle graph to show the number of salespeople in each state.

   Steps: 1. Create a table for the data.

   2. Find the percentage of the whole for each item.

   3. Find the number of degrees for each item.

   4. Draw a circle and draw an angle for each item of data. Label it.

   5. Title the graph.

   Information: Total salespeople 150. Illinois — 65; Iowa — 25; Missouri — 45; Indiana — 15.

   System of Measurement: People

   Computation:

State	IL	IA	MO	IN
Emplys.	65	25	45	15

Illinois: 43% (.4333)   155° (154.8)

Iowa: 17% (.16667)   61° (61.2)

Missouri: 30%   108°

Indiana: 10%   36°

Communicate:

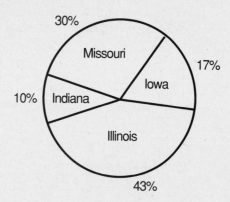

**Percentage of Salespeople
Per State**

JOB SITUATION 2

TASK 1

1. **To Do:** Find the mean overtime wage for the employees in the database.

   Steps: 1. Determine which field or fields contain the information.

   2. Copy the necessary numbers.

   3. Calculate the mean.

   Information: It is found in the field labeled "OT."

   Wages: $13.35 $12.68 $10.05 $12.68 $12.60

   System of Measurement: Money

   Computation:

   mean = (total of the numbers) ÷ (number of items)

   mean = ($13.35 + $12.68 + $10.05 + $12.68 + $12.60) ÷ 5

   mean = $61.36 ÷ 5

   mean = $12.27

   Communicate: The mean overtime hourly wage for these employees is $12.27.

2. **To Do:** Find the mean hourly wage for four employees in the Data Processing Department.

   Steps: 1. Determine which field or fields contain the information.

   2. Copy the necessary numbers.

   3. Calculate the mean.

   Information: It is found in the field labeled "Wage."

   Wages: $8.45 $12.45 $7.90 $8.25

   System of Measurement: Money

Computation:

mean = (total of the numbers) ÷ (number of items)

mean = $37.05 ÷ 4

mean = $9.26

Communicate: The mean hourly wage for these employees is $9.26.

3. **To Do:** Find the mean hourly wage for five employees in the Mailroom.

   Steps: 1. Determine which field or fields contain the information.

   2. Copy the necessary numbers.

   3. Calculate the mean.

   Information: It is found in the field labeled "Wage."

   Wages: $6.70 $5.75 $4.95 $5.00 $6.25

   System of Measurement: Money

   Computation:

   mean = (total of the numbers) ÷ (number of items)

   mean = $28.65 ÷ 5

   mean = $5.73

   Communicate: The mean hourly wage for these employees is $5.73.

4. **To Do:** Find the mean hourly wage for six Copy Center employees in the database.

   Steps: 1. Determine which field or fields contain the information.

   2. Copy the necessary numbers.

   3. Calculate the mean.

   Information: It is found in the field labeled "Wage."

   Wages: $8.90 $7.75 $6.50 $5.90 $8.00 $7.50

   System of Measurement: Money

   Computation:

   mean = (total of the numbers) ÷ (number of items)

   mean = $44.55 ÷ 6

   mean = $7.43

   Communicate: The mean hourly wage for these employees is $7.43.

5. **To Do:** Find the mean hourly wage for six office assistants in the database.

   Steps: 1. Determine which field or fields contain the information.

   2. Copy the necessary numbers.

   3. Calculate the mean.

   Information: It is found in the field labeled "Wage."

   Wages: $9.60 $8.75 $9.00 $8.00 $9.00 $8.25

   System of Measurement: Money

   Computation:

   mean = (total of the numbers) ÷ (number of items)

   mean = $52.60 ÷ 6

   mean = $8.77

   Communicate: The mean hourly wage for these employees is $8.77.

TASK 2

6. **To Do:** Find the range, mode, and median number of employees in the Mailroom.

   Steps: 1. Determine which field or fields contain the information you need.

   2. Copy the necessary numbers, arranging them in order from least to greatest.

   3. Determine the range, mode, and median. ⎰ ⎱

   Information: It is found in the fields labeled by month. Arranged in order: 3 3 3 3 3 4 4 4 4 4 4 4

   System of Measurement: People

   Computation:

   range: 4 - 3 = 1

   mode: 3 3 3 3 3 4 4 4 4 4 4 4

   The most common is 4.

   median: 3 3 3 3 3 4 4 4 4 4 4 4

   4 + 4 = 8 ÷ 2 = 4

   Communicate:

   The range is 1 employee. The mode for the number of employees is 4. The median number of employees is 4.

7. **To Do:** Find the range, mode, and median number of employees in the Data Processing Department.

   Steps: 1. Determine which field or fields contain the information you need.

   2. Copy the necessary numbers, arranging them in order from least to greatest.

   3. Determine the range, mode, and median.

   Information: It is found in the fields labeled by month. Arranged in order: 6 7 7 7 7 7 8 8 9 9 9 9

   System of Measurement: People

   Computation:

   range: 9 - 6 = 3

   mode: The most common is 7.

   median: 7.5

   Communicate:

   The range is 3 employees. The mode for the number of employees is 7. The median number of employees is 7.5.

8. **To Do:** Find the range, mode, and median number of employees in the Sales Department.

   Steps: 1. Determine which field or fields contain the information you need.

   2. Copy the necessary numbers, arranging them in order from least to greatest.

   3. Determine the range, mode, and median.

   Information: It is found in the fields labeled by month. Arranged in order: 22 22 22 22 24 24 25 25 25 26 26 26

System of Measurement: People

Computation:

> range: 26 - 22 = 4
>
> mode: The most common is 22.
>
> median: 24.5

Communicate:

The range is 4 employees. The mode for the number of employees is 22. The median number of employees is 24.5.

9. **To Do:** Find the range, mode, and median number of employees in the Accounting Department.

Steps: 1. Determine which field or fields contain the information you need.

2. Copy the necessary numbers, arranging them in order from least to greatest.

3. Determine the range, mode, and median.

Information: It is found in the fields labeled by month . Arranged in order: 5 5 5 6 7 7 9 10 10 11 12 13

System of Measurement: People

Computation:

> range: 13 - 5 = 8
>
> mode: The most common is 5.
>
> median: 8

Communicate:

The range is 8 employees. The mode for the number of employees is 5. The median number of employees is 8.